Concise Metric Conversion Tables

Compiled by Ann Sleeper

DOLPHIN BOOKS

Doubleday & Company, Inc.
Garden City, New York
1980

No copyright is claimed on the tables in this book, which were taken from publications of the U. S. Department of Commerce/National Bureau of Standards, U. S. Government Printing Office.

ISBN: 0-385-14044-4
Library of Congress Catalog Card Number 79-7701

Contents

Part I

INTRODUCTION: Importance of the Metric System

Why will this book be helpful to you?

Every other industrial nation in the world has officially adopted or committed itself to the use of the Metric System. The great majority of American businessmen and educators believes that the use of this system is in our best interest. In fact, American industry is already making use of it. Current international negotiations are beginning to establish engineering standards for world-wide use. These cover an enormous range—from manufacturing specifications to highway signs to building codes. In a world of growing interdependence, America's industry should be based on international standards—metric standards.

The United States has in fact taken recent steps towards conversion to the Metric System. The Metric Conversion Act of 1975, Public Law 94-168, established the United States Metric Board to coordinate the voluntary conversion to metric measures. This conversion is not mandatory, but it does provide a very necessary framework of timetables and programs for those industries, businesses, and schools that wish to make the change. The United States Metric Board is also conducting a broad public-information program. And almost every State Education Board now requires the metric system to be taught in schools. If mandatory conversion does take place, individuals and organizations will have a well-planned head start.

How Will the New System Affect You?

For most people, the effort involved will be small. The new numbering system will have to be learned. But the system we use now, the Customary System, is one of the most difficult, illogical, confusing and antiquated ever developed. And we would be switching to the simplest, based completely on decimalization, or multiples of ten.

Those people who work with any kind of tool will have to adapt to new sizes. At first during the transition period, these tools will be manufactured in both Metric and Customary sizes until all workers become completely familiar to Metric sizes. The majority of manufacturers are willing to make the switch—by both making new tools and adapting the old ones to the new system. These steps, along with training personnel to use the new devices, will require from five to ten years once the process is officially begun.

Metric System vs. Customary System

There are at present two systems of measurement and weight: the U.S. Customary System and the International (Metric) System.

The Customary System was inherited from the former British Imperial System when the U.S. was still a colony. (Britain has long since abandoned the Imperial System in favor of metric.) Customary measures became established here mainly through use. The Customary units include: inch, foot, yard, mile, pint, quart, gallon, bushel, ounce, pound, and degree Fahrenheit.

The Metric System has now been incorporated into the scientists' new "Système International d'Unités" (International System), a more complete scientific system. All metric measures, other than weight, are based on the meter and its multiples of ten. Because of the simplicity of this decimal system, its use has been steadily increasing in use over the last hundred years.

N.B. Several units of measure remain unchanged between the two systems. *Time* will still be measured in hours, minutes, and seconds. *Electric energy* remains in watts and *light* in lumens. And *money*, already based on the decimal system, will still be counted in dollars, dimes, and cents.

A Brief History of the Metric System

The idea for a metric system began with Gabriel Mouton, a Lyons (France) vicar, in 1670. He proposed a decimal system, and this proposal was discussed many times in the next hundred years. In the 1700s, France's weights and measures were in a chaotic state, and therefore in need of a uniform system. The French Revolution provided the opportunity for the new system to be acted on. In 1790, Charles-Maurice de Talleyrand presented a proposal drawn up by scientists to the National Assembly.

The unit of length in the new system was the meter,

to be equal to a ten-millionth part of the quarter-meridian (North Pole to the Equator) passing through Paris. Some part of that meter cubed was to be the capacity standard which, filled with water, would provide a standard of mass. The aim was thus to interrelate the units for three quantities: length, capacity, and mass.

The metric system was adopted in a bill passed by the French Republic in April 1795. Except for small details of modification, this system today remains the same as it was then presented.

Napoleon is credited for spreading the Metric System in Europe through his military conquests. Though for a long time he allowed both the new system and the old French units to be used, the resulting confusion prompted the French Parliament finally to pass a law in 1840 permitting only the Metric System to be used in public commerce.

In the United States, the Metric System received specific legislative sanction by Congress in the "Law of 1866," and its use has been steadily increasing ever since.

Contents of This Book

In this one volume we have tried to provide the information you need to know and use in order to become familiar with the International (Metric) System.

Part I: Following this Introduction in Part I, you will find Metric Basic Units, where the units of the Metric System are defined, along with the prefixes that modify them by factors of ten. Part I also includes Everyday Units of Measurement, which lists these units for both the Customary and Metric Systems, grouped by category (length, weight, liquid and dry volume, area, time, and temperature).

Part II: The main body of text is Part II, a collection of 54 Conversion Tables for the most commonly used units. With these tables, you have no calculating to do, as the equivalent values for each system are given for every number, starting with one and going sometimes as high as 50,000. Again, these are grouped by category (length, area, weight, liquid and dry volume, and temperature). The tables in the first half convert from the Customary System to the International (Metric) System. Those in the second half convert from Metric to Customary.

Part III: Finally, in Part III, some additional reference is provided. Approximate Metric Conversion Factors gives the numbers to multiply by in order to convert from one unit to another, not only between the two systems but also within the Customary System itself. Here, you will find even the relatively uncommon units of weights and measures.

Part III also includes a Buying Guide, which should prove helpful for travelers in countries where clothing is in metric and a conversion table might not indicate what the French equivalent is to the American size 16 shirt.

A Bibliography concludes the book, listing the publications used in compiling this book, and those the reader may wish to look at for additional information.

METRIC BASIC UNITS

Six units have been adopted to serve as the base for the International System:

Length:	meter
Mass:	kilogram
Time:	second
Electric current:	ampere
Thermodynamic temperature:	degree Kelven
Light intensity:	candela

The basic metric units treated in this book are for LENGTH, MASS, AREA, VOLUME (liquid and dry), and TEMPERATURE.

The *meter* is the unit on which all metric standards and measurements of length, area, and volume are based. The *kilogram*, the basic unit of mass, is independently defined.

Definition of Units

Length

A *meter* is a unit of length equal to 1,650,763.73 wavelengths in a vacuum of the orange-red radiation of krypton 86.

Mass

A *kilogram* is a unit of mass equal to the mass of the International Prototype Kilogram, the mass of a particular platinum-iridium standard.

Capacity, or Volume

A *cubic meter* is a unit of volume equal to a cube the edges of which are 1 meter. A *liter* is a unit of volume equal to a cubic decimeter.

Area

A *square meter* is a unit of area equal to the area of a square the sides of which are 1 meter. A *hectare* is one ten thousandth (1/10,000) of a square meter.

Prefixes

Prefixes indicate multiples or submultiples of the metric units. The most commonly used prefixes, and the multiplication factors they indicate, are:

Prefix:	Multiplication factor
kilo:	1000 (one thousand)
centi:	0.01 (one hundredth)
milli:	0.001 (one thousandth)

Thus, the term kilogram is 1,000 grams; a centigram is 1/100 of a gram; and a milligram is 1/1,000 of a gram.

Other Metric Measurements

Temperature is measured in degrees Celsius. Prefixes are not commonly used. The Metric measurements for *time* and *money* are the same as the Customary measurements.

EVERYDAY UNITS OF MEASUREMENT

The units of Metric and Customary measure given below are not equivalents, except in the case of time, for which the Metric and Customary units are identical.

Unit of Measure	*The Metric System*	*The Customary System*
Length	millimeter	inch
	centimeter	foot
	meter	yard
	kilometer	mile
Weight	gram	ounce
	kilogram	pound
	metric ton (1,000 kilograms)	ton
Volume (liquid)	milliliter	ounce
	liter	cup
		pint
		quart
		gallon
Volume (dry)	cubic millimeters	cubic inches
	cubic centimeters	cubic feet
	cubic meters	cubic yards
	liters	pecks
		bushels
Area	square millimeters	square inches
	square centimeters	square feet
	square meters	square yards
	square kilometers	square miles
	hectares	acres
Time	second	second
	minute	minute
	hour	hour
	day	day
Temperature	degree Celsius	degree Fahrenheit

Part II Conversion Tables

CUSTOMARY TO METRIC

Length

Table 1 Fractions of an Inch to Millimeters

1 inch = 25.40 millimeters

in.	mm.	in.	mm.
1/64	.40	33/64	13.10
1/32	.79	17/32	13.49
3/64	1.19	35/64	13.89
1/16	1.59	9/16	14.29
5/64	1.98	37/64	14.68
3/32	2.38	19/32	15.08
7/64	2.78	39/64	15.48
1/8	3.18	5/8	15.88
9/64	3.57	41/64	16.27
5/32	3.97	21/32	16.67
11/64	4.37	43/64	17.07
3/16	4.76	11/16	17.46
13/64	5.16	45/64	17.86
7/32	5.56	23/32	18.26
15/64	5.95	47/64	18.65
1/4	6.35	3/4	19.05
17/64	6.75	49/64	19.45
9/32	7.14	25/32	19.84
19/64	7.54	51/64	20.24
5/16	7.94	13/16	20.64
21/64	8.33	53/64	21.03
11/32	8.73	27/32	21.43
23/64	9.13	55/64	21.83
3/8	9.53	7/8	22.23
25/64	9.92	57/64	22.62
13/32	10.32	29/32	23.02
27/64	10.72	59/64	23.42
7/16	11.11	15/16	23.81
29/64	11.51	61/64	24.21
15/32	11.91	31/32	24.61
31/64	12.30	63/64	25.00
1/2	12.70		

Table 2 Feet, Inches to Centimeters

1 inch = 2.54 centimeters

1 foot = 30.48 centimeters

ft.	in.	cm.	ft.	in.	cm.
	1	2.54	2	8	81.28
	2	5.08	2	9	83.82
	3	7.62	2	10	86.36
	4	10.16	2	11	88.90
	5	12.70	3		91.44
	6	15.24	3	1	93.98
	7	17.78	3	2	96.52
	8	20.32	3	3	99.06
	9	22.86	3	4	101.60
	10	25.40	3	5	104.14
	11	27.94	3	6	106.68
1		30.48	3	7	109.22
1	1	33.02	3	8	111.76
1	2	35.56	3	9	114.30
1	3	38.10	3	10	116.84
1	4	40.64	3	11	119.38
1	5	43.18	4		121.92
1	6	45.72	4	1	124.46
1	7	48.26	4	2	127.00
1	8	50.80	4	3	129.54
1	9	53.34	4	4	132.08
1	10	55.88	4	5	134.62
1	11	58.42	4	6	137.16
2		60.96	4	7	139.70
2	1	63.50	4	8	142.24
2	2	66.04	4	9	144.78
2	3	68.58	4	10	147.32
2	4	71.12	4	11	149.86
2	5	73.66	5		152.40
2	6	76.20	5	1	154.94
2	7	78.74	5	2	157.48

Table 2 *(continued)*

ft.	in.	cm.	ft.	in.	cm.
5	3	160.02	5	8	172.72
5	4	162.56	5	9	175.26
5	5	165.10	5	10	177.80
5	6	167.64	5	11	180.34
5	7	170.18	6		182.88

Table 3 Yards, Feet to Meters

1 foot = .3048 meter

1 yard = .9144 meter

yd.	ft.	m.	yd.	ft.	m.
	1	.3048	10	2	9.7536
	2	.6096	11		10.0584
1		.9144	11	1	10.3632
1	1	1.2192	11	2	10.6680
1	2	1.5240	12		10.9728
2		1.8288	12	1	11.2776
2	1	2.1336	12	2	11.5824
2	2	2.4384	13		11.8872
3		2.7432	13	1	12.1920
3	1	3.0480	13	2	12.4968
3	2	3.3528	14		12.8016
4		3.6576	14	1	13.1064
4	1	3.9624	14	2	13.4112
4	2	4.2672	15		13.7160
5		4.5720	15	1	14.0208
5	1	4.8768	15	2	14.3256
5	2	5.1816	16		14.6304
6		5.4864	16	1	14.9352
6	1	5.7912	16	2	15.2400
6	2	6.0960	17		15.5448
7		6.4008	17	1	15.8496
7	1	6.7056	17	2	16.1544
7	2	7.0104	18		16.4592
8		7.3152	18	1	16.7640
8	1	7.6200	18	2	17.0688
8	2	7.9248	19		17.3736
9		8.2296	19	1	17.6784
9	1	8.5344	19	2	17.9832
9	2	8.8392	20		18.2880
10		9.1440	20	1	18.5928
10	1	9.4488	20	2	18.8976

Table 3 *(continued)*

yd.	ft.	m.	yd.	ft.	m.
21		19.2024	32	1	29.5656
21	1	19.5072	32	2	29.8704
21	2	19.8120	33		30.1752
22		20.1168	33	1	30.4800
22	1	20.4216	33	2	30.7848
22	2	20.7264	34		31.0896
23		21.0312	34	1	31.3944
23	1	21.3360	34	2	31.6992
23	2	21.6408	35		32.0040
24		21.9456	35	1	32.3088
24	1	22.2504	35	2	32.6136
24	2	22.5552	36		32.9184
25		22.8600	36	1	33.2232
25	1	23.1648	36	2	33.5280
25	2	23.4696	37		33.8328
26		23.7744	37	1	34.1376
26	1	24.0792	37	2	34.4424
26	2	24.3840	38		34.7472
27		24.6888	38	1	35.0520
27	1	24.9936	38	2	35.3568
27	2	25.2984	39		35.6616
28		25.6032	39	1	35.9664
28	1	25.9080	39	2	36.2712
28	2	26.2128	40		36.5760
29		26.5176	40	1	36.8808
29	1	26.8224	40	2	37.1856
29	2	27.1272	41		37.4904
30		27.4320	41	1	37.7952
30	1	27.7368	41	2	38.1000
30	2	28.0416	42		38.4048
31		28.3464	42	1	38.7096
31	1	28.6512	42	2	39.0144
31	2	28.9560	43		39.3192
32		29.2608	43	1	39.6240

Table 3 *(continued)*

yd.	ft.	m.
43	2	39.9288
44		40.2336
44	1	40.5384
44	2	40.8432
45		41.1480
45	1	41.4528
45	2	41.7576
46		42.0624
46	1	42.3672
46	2	42.6720
47		42.9768
47	1	43.2816
47	2	43.5864
48		43.8912
48	1	44.1960
48	2	44.5008
49		44.8056
49	1	45.1104
49	2	45.4152
50		45.7200
50	1	46.0248
50	2	46.3296
51		46.6344
51	1	46.9392
51	2	47.2440
52		47.5488
52	1	47.8536
52	2	48.1584
53		48.4632
53	1	48.7680
53	2	49.0728
54		49.3776
54	1	49.6824
54	2	49.9872

yd.	ft.	m.
55		50.2920
55	1	50.5968
55	2	50.9016
56		51.2064
56	1	51.5112
56	2	51.8160
57		52.1208
57	1	52.4256
57	2	52.7304
58		53.0352
58	1	53.3400
58	2	53.6448
59		53.9496
59	1	54.2544
59	2	54.5592
60		54.8640
60	1	55.1688
60	2	55.4736
61		55.7784
61	1	56.0832
61	2	56.3880
62		56.6928
62	1	56.9976
62	2	57.3024
63		57.6072
63	1	57.9120
63	2	58.2168
64		58.5126
64	1	58.8264
64	2	59.1312
65		59.4360
65	1	59.7408
65	2	60.0456
66		60.3504

Table 3 *(continued)*

yd.	ft.	m.	yd.	ft.	m.
66	1	60.6552	77	2	71.0184
66	2	60.9600	78		71.3232
67		61.2648	78	1	71.6280
67	1	61.5696	78	2	71.9328
67	2	61.8744	79		72.2376
68		62.1792	79	1	72.5424
68	1	62.4840	79	2	72.8472
68	2	62.7888	80		73.1520
69		63.0936	80	1	73.4568
69	1	63.3984	80	2	73.7616
69	2	63.7032	81		74.0664
70		64.0080	81	1	74.3712
70	1	64.3128	81	2	74.6760
70	2	64.6176	82		74.9808
71		64.9224	82	1	75.2856
71	1	65.2272	82	2	75.6904
71	2	65.5320	83		75.8952
72		65.8368	83	1	76.2000
72	1	66.1416	83	2	76.5048
72	2	66.4464	84		76.8096
73		66.7512	84	1	77.1144
73	1	67.0560	84	2	77.4192
73	2	67.3608	85		77.7240
74		67.6656	85	1	78.0288
74	1	67.9704	85	2	78.3336
74	2	68.2752	86		78.6384
75		68.5800	86	1	78.9432
75	1	68.8848	86	2	79.2480
75	2	69.1896	87		79.5528
76		69.4944	87	1	79.8576
76	1	69.7992	87	2	80.1624
76	2	70.1040	88		80.4672
77		70.4088	88	1	80.7720
77	1	70.7136	88	2	81.0768

Table 3 *(continued)*

yd.	ft.	m.	yd.	ft.	m.
89		81.3816	94	2	86.5632
89	1	81.6864	95		86.8680
89	2	81.9912	95	1	87.1728
90		82.2960	95	2	87.4776
90	1	82.6008	96		87.7824
90	2	82.9056	96	1	88.0872
91		83.2104	96	2	88.3920
91	1	83.5152	97		88.6968
91	2	83.8200	97	1	89.0016
92		84.1248	97	2	89.3064
92	1	84.4296	98		89.6112
92	2	84.7344	98	1	89.9160
93		85.0392	98	2	90.2208
93	1	85.3440	99		90.5256
93	2	85.6488	99	1	90.8304
94		85.9536	99	2	91.1352
94	1	86.2584	100		91.4400

Table 4 Decimals of a Mile to Feet to Meters to Kilometers

1 mile = 5280 feet

1 foot = .3048 meter

1 mile = 1.609 kilometers

decimals of mi.	ft.	m.	km.
0.1	528	160.9344	.1609
0.2	1056	321.8688	.3219
0.25	1320	402.3360	.4023
0.3	1584	482.8032	.4828
0.4	2112	643.7376	.6437
0.5	2640	804.6720	.8047
0.6	3168	965.6064	.9656
0.7	3696	1126.5408	1.1265
0.75	3960	1207.0080	1.2070
0.8	4224	1287.4752	1.2875
0.9	4752	1448.4096	1.4484
1.0	5280	1609.3440	1.6093

Table 5 Miles to Kilometers

1 mile = 1.609 kilometers

mi.	km.	mi.	km.	mi.	km.
1	1.6093	33	53.1084	65	104.6074
2	3.2187	34	54.7177	66	106.2167
3	4.8280	35	56.3270	67	107.8260
4	6.4374	36	57.9364	68	109.4354
5	8.0467	37	59.5457	69	111.0447
6	9.6561	38	61.1551	70	112.6541
7	11.2654	39	62.7644	71	114.2634
8	12.8748	40	64.3738	72	115.8728
9	14.4841	41	65.9831	73	117.4821
10	16.0934	42	67.5924	74	119.0915
11	17.7028	43	69.2018	75	120.7008
12	19.3121	44	70.8111	76	122.3101
13	20.9215	45	72.4205	77	123.9195
14	22.5308	46	74.0298	78	125.5288
15	24.1402	47	75.6392	79	127.1382
16	25.7495	48	77.2485	80	128.7475
17	27.3588	49	78.8579	81	130.3569
18	28.9682	50	80.4672	82	131.9662
19	30.5775	51	82.0765	83	133.5756
20	32.1869	52	83.6859	84	135.1849
21	33.7962	53	85.2952	85	136.7942
22	35.4056	54	86.9046	86	138.4036
23	37.0149	55	88.5139	87	140.0129
24	38.6243	56	90.1233	88	141.6223
25	40.2336	57	91.7326	89	143.2316
26	41.8429	58	93.3420	90	144.8410
27	43.4523	59	94.9513	91	146.4503
28	45.0616	60	96.5606	92	148.0596
29	46.6710	61	98.1700	93	149.6690
30	48.2803	62	99.7793	94	151.2783
31	49.8897	63	101.3887	95	152.8877
32	51.4990	64	102.9980	96	154.4970

Table 5 *(continued)*

mi.	km.	mi.	km.	mi.	km.
97	156.1064	131	210.8241	165	265.5418
98	157.7157	132	212.4334	166	267.1511
99	159.3251	133	214.0428	167	268.7604
100	160.9344	134	215.6521	168	270.3698
101	162.5437	135	217.2614	169	271.9791
102	164.1531	136	218.8708	170	273.5885
103	165.7624	137	220.4801	171	275.1978
104	167.3718	138	222.0895	172	276.8072
105	168.9811	139	223.6988	173	278.4165
106	170.5905	140	225.3082	174	280.0259
107	172.1998	141	226.9175	175	281.6352
108	173.8092	142	228.5268	176	283.2445
109	175.4185	143	230.1362	177	284.8539
110	177.0278	144	231.7455	178	286.4632
111	178.6372	145	233.3549	179	288.0726
112	180.2465	146	234.9642	180	289.6819
113	181.8559	147	236.5736	181	291.2913
114	183.4652	148	238.1829	182	292.9006
115	185.0746	149	239.7923	183	294.5100
116	186.6839	150	241.4016	184	296.1193
117	188.2932	151	243.0109	185	297.7286
118	189.9026	152	244.6203	186	299.3380
119	191.5119	153	246.2296	187	300.9473
120	193.1213	154	247.8390	188	302.5567
121	194.7306	155	249.4483	189	304.1660
122	196.3400	156	251.0577	190	305.7754
123	197.9493	157	252.6670	191	307.3847
124	199.5587	158	254.2764	192	308.9940
125	201.1680	159	255.8857	193	310.6034
126	202.7773	160	257.4950	194	312.2127
127	204.3867	161	259.1044	195	313.8221
128	205.9960	162	260.7137	196	315.4314
129	207.6054	163	262.3231	197	317.0408
130	209.2147	164	263.9324	198	318.6501

Table 5 *(continued)*

mi.	km.	mi.	km.	mi.	km.
199	320.2595	233	374.9771	267	429.6948
200	321.8688	234	376.5865	268	431.3042
201	323.4781	235	378.1958	269	432.9135
202	325.0875	236	379.8052	270	434.5229
203	326.6968	237	381.4145	271	436.1322
204	328.3062	238	383.0239	272	437.7416
205	329.9155	239	384.6332	273	439.3509
206	331.5249	240	386.2426	274	440.9603
207	333.1342	241	387.8519	275	442.5696
208	334.7436	242	389.4612	276	444.1789
209	336.3529	243	391.0706	277	445.7883
210	337.9622	244	392.6799	278	447.3976
211	339.5716	245	394.2893	279	449.0070
212	341.1809	246	395.8986	280	450.6163
213	342.7903	247	397.5080	281	452.2257
214	344.3996	248	399.1173	282	453.8350
215	346.0090	249	400.7267	283	455.4444
216	347.6183	250	402.3360	284	457.0537
217	349.2276	251	403.9453	285	458.6630
218	350.8370	252	405.5547	286	460.2724
219	352.4463	253	407.1640	287	461.8817
220	354.0557	254	408.7734	288	463.4911
221	355.6650	255	410.3827	289	465.1004
222	357.2744	256	411.9921	290	466.7098
223	358.8837	257	413.6014	291	468.3191
224	360.4931	258	415.2108	292	469.9284
225	362.1024	259	416.8201	293	471.5378
226	363.7117	260	418.4294	294	473.1471
227	365.3211	261	420.0388	295	474.7565
228	366.9304	262	421.6481	296	476.3658
229	368.5398	263	423.2575	297	477.9752
230	370.1491	264	424.8668	298	479.5845
231	371.7585	265	426.4762	299	481.1939
232	373.3678	266	428.0855	300	482.8032

Table 5 *(continued)*

mi.	km.	mi.	km.	mi.	km.
301	484.4125	335	539.1302	369	593.8479
302	486.0219	336	540.7396	370	595.4573
303	487.6312	337	542.3489	371	597.0666
304	489.2406	338	543.9583	372	598.6760
305	490.8499	339	545.5676	373	600.2853
306	492.4593	340	547.1770	374	601.8947
307	494.0686	341	548.7863	375	603.5040
308	495.6780	342	550.3956	376	605.1133
309	497.2873	343	552.0050	377	606.7227
310	498.8966	344	553.6143	378	608.3320
311	500.5060	345	555.2237	379	609.9414
312	502.1153	346	556.8330	380	611.5507
313	503.7247	347	558.4424	381	613.1601
314	505.3340	348	560.0517	382	614.7694
315	506.9434	349	561.6611	383	616.3788
316	508.5527	350	563.2704	384	617.9881
317	510.1620	351	564.8797	385	619.5974
318	511.7714	352	566.4891	386	621.2068
319	513.3807	353	568.0984	387	622.8161
320	514.9901	354	569.7078	388	624.4255
321	516.5994	355	571.3171	389	626.0348
322	518.2088	356	572.9265	390	627.6442
323	519.8181	357	574.5358	391	629.2535
324	521.4275	358	576.1451	392	630.8628
325	523.0368	359	577.7545	393	632.4722
326	524.6461	360	579.3638	394	634.0815
327	526.2555	361	580.9732	395	635.6909
328	527.8648	362	582.5825	396	637.3002
329	529.4742	363	584.1919	397	638.9096
330	531.0835	364	585.8012	398	640.5189
331	532.6929	365	587.4106	399	642.1283
332	534.3022	366	589.0199	400	643.7376
333	535.9116	367	590.6292	401	645.3469
334	537.5209	368	592.2386	402	646.9563

Table 5 *(continued)*

mi.	km.	mi.	km.	mi.	km.
403	648.5656	437	703.2833	471	758.0010
404	650.1750	438	704.8927	472	759.6104
405	651.7843	439	706.5020	473	761.2197
406	653.3937	440	708.1114	474	762.8291
407	655.0030	441	709.7207	475	764.4384
408	656.6124	442	711.3300	476	766.0477
409	658.2217	443	712.9394	477	767.6571
410	659.8310	444	714.5487	478	769.2664
411	661.4404	445	716.1581	479	770.8758
412	663.0497	446	717.7674	480	772.4851
413	664.6591	447	719.3768	481	774.0945
414	666.2684	448	720.9861	482	775.7038
415	667.8778	449	722.5955	483	777.3131
416	669.4871	450	724.2048	484	778.9225
417	671.0964	451	725.8141	485	780.5318
418	672.7058	452	727.4235	486	782.1412
419	674.3151	453	729.0328	487	783.7505
420	675.9245	454	730.6422	488	785.3599
421	677.5338	455	732.2515	489	786.9692
422	679.1432	456	733.8609	490	788.5786
423	680.7525	457	735.4702	491	790.1879
424	682.3619	458	737.0796	492	791.7972
425	683.9712	459	738.6889	493	793.4066
426	685.5805	460	740.2982	494	795.0159
427	687.1899	461	741.9076	495	796.6253
428	688.7992	462	743.5169	496	798.2346
429	690.4086	463	745.1263	497	799.8440
430	692.0179	464	746.7356	498	801.4533
431	693.6273	465	748.3450	499	803.0627
432	695.2366	466	749.9543	500	804.6720
433	696.8459	467	751.5636	510	820.7654
434	698.4553	468	753.1730	520	836.8589
435	700.0646	469	754.7823	530	852.9523
436	701.6740	470	756.3917	540	869.0458

Table 5 *(continued)*

mi.	km.	mi.	km.	mi.	km.
550	885.1392	710	1142.6342	860	1384.0359
560	901.2326	720	1158.7277	870	1400.1293
570	917.3261	730	1174.8211	880	1416.2227
580	933.4195	740	1190.9146	890	1432.3162
590	949.5130	750	1207.0080	900	1448.4096
600	965.6064	760	1223.1014	910	1464.5030
610	981.6998	770	1239.1949	920	1480.5965
620	997.7933	780	1255.2883	930	1496.6899
630	1013.8867	790	1271.3818	940	1512.7834
640	1029.9802	800	1287.4752	950	1528.8768
650	1046.0736	810	1303.5686	960	1544.9702
660	1062.1670	820	1319.6621	970	1561.0637
670	1078.2605	830	1335.7555	980	1577.1571
680	1094.3539	840	1351.8490	990	1593.2506
690	1110.4474	850	1367.9424	1000	1609.3440
700	1126.5408				

Table 6 Knots (Nautical Miles) to Kilometers

1 knot = 1.852 kilometers

k.	km.	k.	km.	k.	km.
1	1.852	33	61.116	65	120.380
2	3.704	34	62.968	66	122.232
3	5.556	35	64.820	67	124.084
4	7.408	36	66.672	68	125.936
5	9.260	37	68.524	69	127.788
6	11.112	38	70.376	70	129.640
7	12.964	39	72.228	71	131.492
8	14.816	40	74.080	72	133.344
9	16.668	41	75.932	73	135.196
10	18.520	42	77.784	74	137.048
11	20.372	43	79.636	75	138.900
12	22.224	44	81.488	76	140.752
13	24.076	45	83.340	77	142.604
14	25.928	46	85.192	78	144.456
15	27.780	47	87.044	79	146.308
16	29.632	48	88.896	80	148.160
17	31.484	49	90.748	81	150.012
18	33.336	50	92.600	82	151.864
19	35.188	51	94.452	83	153.716
20	37.040	52	96.304	84	155.568
21	38.892	53	98.156	85	157.420
22	40.744	54	100.008	86	159.272
23	42.596	55	101.860	87	161.124
24	44.448	56	103.712	88	162.976
25	46.300	57	105.564	89	164.828
26	48.152	58	107.416	90	166.680
27	50.004	59	109.268	91	168.532
28	51.856	60	111.120	92	170.384
29	53.708	61	112.972	93	172.236
30	55.560	62	114.824	94	174.088
31	57.412	63	116.676	95	175.940
32	59.264	64	118.528	96	177.792

Table 6 *(continued)*

k.	km.	k.	km.	k.	km.
97	179.644	132	244.464	167	309.284
98	181.496	133	246.316	168	311.136
99	183.348	134	248.168	169	312.988
100	185.200	135	250.020	170	314.840
101	187.052	136	251.872	171	316.692
102	188.904	137	253.724	172	318.544
103	190.756	138	255.576	173	320.396
104	192.608	139	257.428	174	322.248
105	194.460	140	259.280	175	324.100
106	196.312	141	261.132	176	325.952
107	198.164	142	262.984	177	327.804
108	200.016	143	264.836	178	329.656
109	201.868	144	266.688	179	331.508
110	203.720	145	268.540	180	333.360
111	205.572	146	270.392	181	335.212
112	207.424	147	272.244	182	337.064
113	209.276	148	274.096	183	338.916
114	211.128	149	275.948	184	340.768
115	212.980	150	277.800	185	342.620
116	214.832	151	279.652	186	344.472
117	216.684	152	281.504	187	346.324
118	218.536	153	283.356	188	348.176
119	220.388	154	285.208	189	350.028
120	222.240	155	287.060	190	351.880
121	224.092	156	288.912	191	353.732
122	225.944	157	290.764	192	355.584
123	227.796	158	292.616	193	357.436
124	229.648	159	294.468	194	359.288
125	231.500	160	296.320	195	361.140
126	233.352	161	298.172	196	362.992
127	235.204	162	300.024	197	364.844
128	237.056	163	301.876	198	366.696
129	238.908	164	303.728	199	368.548
130	240.760	165	305.580	200	370.400
131	242.612	166	307.432		

Table 7 Miles per Gallon to Kilometers per Liter

1 mile per gallon = .43 kilometer per liter

mi./gal.	km./l.	mi./gal.	km./l.
1	.43	33	14.0
2	.85	34	14.5
3	1.28	35	14.9
4	1.70	36	15.3
5	2.13	37	15.7
6	2.55	38	16.2
7	2.98	39	16.6
8	3.40	40	17.0
9	3.83	41	17.4
10	4.25	42	17.9
11	4.7	43	18.3
12	5.1	44	18.7
13	5.5	45	19.1
14	6.0	46	19.6
15	6.4	47	20.0
16	6.8	48	20.4
17	7.2	49	20.8
18	7.7	50	21.3
19	8.1	51	21.7
20	8.5	52	22.1
21	8.9	53	22.5
22	9.4	54	23.0
23	9.8	55	23.4
24	10.2	56	23.8
25	10.6	57	24.2
26	11.1	58	24.7
27	11.5	59	25.1
28	11.9	60	25.5
29	12.3	61	25.9
30	12.8	62	26.4
31	13.2	63	26.8
32	13.6	64	27.2

Table 7 *(continued)*

mi./gal.	km./l.	mi./gal.	km./l.
65	27.6	83	35.3
66	28.1	84	35.7
67	28.5	85	36.1
68	28.9	86	36.6
69	29.3	87	37.0
70	29.8	88	37.4
71	30.2	89	37.8
72	30.6	90	38.3
73	31.0	91	38.7
74	31.5	92	39.1
75	31.9	93	39.5
76	32.3	94	40.0
77	32.7	95	40.4
78	33.2	96	40.8
79	33.6	97	41.2
80	34.0	98	41.7
81	34.4	99	42.1
82	34.9	100	42.5

Area

Table 8 Square Inches to Square Centimeters

1 square inch = 6.4516 square centimeters

in.^{2}	cm.^{2}	in.^{2}	cm.^{2}	in.^{2}	cm.^{2}
0.1	.6452	22	141.9352	54	348.3864
0.2	1.2903	23	148.3868	55	354.8380
0.25	1.6129	24	154.8384	56	361.2896
0.3	1.9355	25	161.2900	57	367.7412
0.4	2.5806	26	167.7416	58	374.1928
0.5	3.2258	27	174.1932	59	380.6444
0.6	3.8710	28	180.6448	60	387.0960
0.7	4.5161	29	187.0964	61	393.5476
0.75	4.8387	30	193.5480	62	399.9992
0.8	5.1613	31	199.9996	63	406.4508
0.9	5.8064	32	206.4512	64	412.9024
1	6.4516	33	212.9028	65	419.3540
2	12.9032	34	219.3544	66	425.8056
3	19.3548	35	225.8060	67	432.2572
4	25.8064	36	232.2576	68	438.7088
5	32.2580	37	238.7092	69	445.1604
6	38.7096	38	245.1608	70	451.6120
7	45.1612	39	251.6124	71	458.0636
8	51.6128	40	258.0640	72	464.5152
9	58.0644	41	264.5156	73	470.9668
10	64.5160	42	270.9672	74	477.4184
11	70.9676	43	277.4188	75	483.8700
12	77.4192	44	283.8704	76	490.3216
13	83.8708	45	290.3220	77	496.7732
14	90.3224	46	296.7736	78	503.2248
15	96.7740	47	303.2252	79	509.6764
16	103.2256	48	309.6768	80	516.1280
17	109.6772	49	316.1284	81	522.5796
18	116.1288	50	322.5800	82	529.0312
19	122.5804	51	329.0316	83	535.4828
20	129.0320	52	335.4832	84	541.9344
21	135.4836	53	341.9348	85	548.3860

Table 8 *(continued)*

in.²	cm.²	in.²	cm.²	in.²	cm.²
86	554.8376	120	774.1920	154	993.5464
87	561.2892	121	780.6436	155	999.9980
88	567.7408	122	787.0952	156	1006.4496
89	574.1924	123	793.5468	157	1012.9012
90	580.6440	124	799.9984	158	1019.3528
91	587.0956	125	806.4500	159	1025.8044
92	593.5472	126	812.9016	160	1032.2560
93	599.9988	127	819.3532	161	1038.7076
94	606.4504	128	825.8048	162	1045.1592
95	612.9020	129	832.2564	163	1051.6108
96	619.3536	130	838.7080	164	1058.0624
97	625.8052	131	845.1596	165	1064.5140
98	632.2568	132	851.6112	166	1070.9656
99	638.7084	133	858.0628	167	1077.4172
100	645.1600	134	864.5144	168	1083.8688
101	651.6116	135	870.9660	169	1090.3204
102	658.0632	136	877.4176	170	1096.7720
103	664.5148	137	883.8692	171	1103.2236
104	670.9664	138	890.3208	172	1109.6752
105	677.4180	139	896.7724	173	1116.1268
106	683.8696	140	903.2240	174	1122.5784
107	690.3212	141	909.6756	175	1129.0300
108	696.7728	142	916.1272	176	1135.4816
109	703.2244	143	922.5788	177	1141.9332
110	709.6760	144	929.0304	178	1148.3848
111	716.1276	145	935.4820	179	1154.8364
112	722.5792	146	941.9336	180	1161.2880
113	729.0308	147	948.3852	181	1167.7396
114	735.4824	148	954.8368	182	1174.1912
115	741.9340	149	961.2884	183	1180.6428
116	748.3856	150	967.7400	184	1187.0944
117	754.8372	151	974.1916	185	1193.5460
118	761.2888	152	980.6432	186	1199.9976
119	767.7404	153	987.0948	187	1206.4492

Table 8 *(continued)*

in.^{2}	cm.^{2}	in.^{2}	cm.^{2}	in.^{2}	cm.^{2}
188	1212.9008	222	1432.2552	256	1651.6096
189	1219.3524	223	1438.7068	257	1658.0612
190	1225.8040	224	1445.1584	258	1664.5128
191	1232.2556	225	1451.6100	259	1670.9644
192	1238.7072	226	1458.0616	260	1677.4160
193	1245.1588	227	1464.5132	261	1683.8676
194	1251.6104	228	1470.9648	262	1690.3192
195	1258.0620	229	1477.4164	263	1696.7708
196	1264.5136	230	1483.8680	264	1703.2224
197	1270.9652	231	1490.3196	265	1709.6740
198	1277.4168	232	1496.7712	266	1716.1256
199	1283.8684	233	1503.2228	267	1722.5772
200	1290.3200	234	1509.6744	268	1729.0288
201	1296.7716	235	1516.1260	269	1735.4804
202	1303.2232	236	1522.5776	270	1741.9320
203	1309.6748	237	1529.0292	271	1748.3836
204	1316.1264	238	1535.4808	272	1754.8352
205	1322.5780	239	1541.9324	273	1761.2868
206	1329.0296	240	1548.3840	274	1767.7384
207	1335.4812	241	1554.8356	275	1774.1900
208	1341.9328	242	1561.2872	276	1780.6416
209	1348.3844	243	1567.7388	277	1787.0932
210	1354.8360	244	1574.1904	278	1793.5448
211	1361.2876	245	1580.6420	279	1799.9964
212	1367.7392	246	1587.0936	280	1806.4480
213	1374.1908	247	1593.5452	281	1812.8996
214	1380.6424	248	1599.9968	282	1819.3512
215	1387.0940	249	1606.4484	283	1825.8028
216	1393.5456	250	1612.9000	284	1832.2544
217	1399.9972	251	1619.3516	285	1838.7060
218	1406.4488	252	1625.8032	286	1845.1576
219	1412.9004	253	1632.2548	287	1851.6092
220	1419.3520	254	1638.7064	288	1858.0608
221	1425.8036	255	1645.1580	289	1864.5124

Table 8 *(continued)*

in.²	cm.²	in.²	cm.²	in.²	cm.²
290	1870.9640	324	2090.3184	358	2309.6728
291	1877.4156	325	2096.7700	359	2316.1244
292	1883.8672	326	2103.2216	360	2322.5760
293	1890.3188	327	2109.6732	361	2329.0276
294	1896.7704	328	2116.1248	362	2335.4792
295	1903.2220	329	2122.5764	363	2341.9308
296	1909.6736	330	2129.0280	364	2348.3824
297	1916.1252	331	2135.4796	365	2354.8340
298	1922.5768	332	2141.9312	366	2361.2856
299	1929.0284	333	2148.3828	367	2367.7372
300	1935.4800	334	2154.8344	368	2374.1888
301	1941.9316	335	2161.2860	369	2380.6404
302	1948.3832	336	2167.7376	370	2387.0920
303	1954.8348	337	2174.1892	371	2393.5436
304	1961.2864	338	2180.6408	372	2399.9952
305	1967.7380	339	2187.0924	373	2406.4468
306	1974.1896	340	2193.5440	374	2412.8984
307	1980.6412	341	2199.9956	375	2419.3500
308	1987.0928	342	2206.4472	376	2425.8016
309	1993.5444	343	2212.8988	377	2432.2532
310	1999.9960	344	2219.3504	378	2438.7048
311	2006.4476	345	2225.8020	379	2445.1564
312	2012.8992	346	2232.2536	380	2451.6080
313	2019.3508	347	2238.7052	381	2458.0596
314	2025.8024	348	2245.1568	382	2464.5112
315	2032.2540	349	2251.6084	383	2470.9628
316	2038.7056	350	2258.0600	384	2477.4144
317	2045.1572	351	2264.5116	385	2483.8660
318	2051.6088	352	2270.9632	386	2490.3176
319	2058.0604	353	2277.4148	387	2496.7692
320	2064.5120	354	2283.8664	388	2503.2208
321	2070.9636	355	2290.3180	389	2509.6724
322	2077.4152	356	2296.7696	390	2516.1240
323	2083.8668	357	2303.2212	391	2522.5756

Table 8 *(continued)*

in.^{2}	cm.^{2}	in.^{2}	cm.^{2}	in.^{2}	cm.^{2}
392	2529.0272	426	2748.3816	460	2967.7360
393	2535.4788	427	2754.8332	461	2974.1876
394	2541.9304	428	2761.2848	462	2980.6392
395	2548.3820	429	2767.7364	463	2987.0908
396	2554.8336	430	2774.1880	464	2993.5424
397	2561.2852	431	2780.6396	465	2999.9940
398	2567.7368	432	2787.0912	466	3006.4456
399	2574.1884	433	2793.5428	467	3012.8972
400	2580.6400	434	2799.9944	468	3019.3488
401	2587.0916	435	2806.4460	469	3025.8004
402	2593.5432	436	2812.8976	470	3032.2520
403	2599.9948	437	2819.3492	471	3038.7036
404	2606.4464	438	2825.8008	472	3045.1552
405	2612.8980	439	2832.2524	473	3051.6068
406	2619.3496	440	2838.7040	474	3058.0584
407	2625.8012	441	2845.1556	475	3064.5100
408	2632.2528	442	2851.6072	476	3070.9616
409	2638.7044	443	2858.0588	477	3077.4132
410	2645.1560	444	2864.5104	478	3083.8648
411	2651.6076	445	2870.9620	479	3090.3164
412	2658.0592	446	2877.4136	480	3096.7680
413	2664.5108	447	2883.8652	481	3103.2196
414	2670.9624	448	2890.3168	482	3109.6712
415	2677.4140	449	2896.7684	483	3116.1228
416	2683.8656	450	2903.2200	484	3122.5744
417	2690.3172	451	2909.6716	485	3129.0260
418	2696.7688	452	2916.1232	486	3135.4776
419	2703.2204	453	2922.5748	487	3141.9292
420	2709.6720	454	2929.0264	488	3148.3808
421	2716.1236	455	2935.4780	489	3154.8324
422	2722.5752	456	2941.9296	490	3161.2840
423	2729.0268	457	2948.3812	491	3167.7356
424	2735.4784	458	2954.8328	492	3174.1872
425	2741.9300	459	2961.2844	493	3180.6388

Table 8 *(continued)*

in.^{2}	cm.^{2}	in.^{2}	cm.^{2}	in.^{2}	cm.^{2}
494	3187.0904	630	4064.5080	820	5290.3120
495	3193.5420	640	4129.0240	830	5354.8280
496	3199.9936	650	4193.5400	840	5419.3440
497	3206.4452	660	4258.0560	850	5483.8600
498	3212.8968	670	4322.5720	860	5548.3760
499	3219.3484	680	4387.0880	870	5612.8920
500	3225.8000	690	4451.6040	880	5677.4080
510	3290.3160	700	4516.1200	890	5741.9240
520	3354.8320	710	4580.6360	900	5806.4400
530	3419.3480	720	4645.1520	910	5870.9560
540	3483.8640	730	4709.6680	920	5935.4720
550	3548.3800	740	4774.1840	930	5999.9880
560	3612.8960	750	4838.7000	940	6064.5040
570	3677.4120	760	4903.2160	950	6129.0200
580	3741.9280	770	4967.7320	960	6193.5360
590	3806.4440	780	5032.2480	970	6258.0520
600	3870.9600	790	5096.7640	980	6322.5680
610	3935.4760	800	5161.2800	990	6387.0840
620	3999.9920	810	5225.7960	1000	6451.6000

Table 9 Square Feet to Square Meters

1 square foot = .0929 square meter

ft.^{2}	m.^{2}	ft.^{2}	m.^{2}	ft.^{2}	m.^{2}
1	0.09290	33	3.06580	65	6.03870
2	0.18581	34	3.15870	66	6.13160
3	0.27871	35	3.25161	67	6.22450
4	0.37161	36	3.34451	68	6.31741
5	0.46452	37	3.43741	69	6.41031
6	0.55742	38	3.53032	70	6.50321
7	0.65032	39	3.62322	71	6.59612
8	0.74322	40	3.71612	72	6.68902
9	0.83613	41	3.80902	73	6.78192
10	0.92903	42	3.90193	74	6.87482
11	1.02193	43	3.99483	75	6.96773
12	1.11484	44	4.08773	76	7.06063
13	1.20774	45	4.18064	77	7.15353
14	1.30064	46	4.27354	78	7.24644
15	1.39355	47	4.36644	79	7.33934
16	1.48645	48	4.45935	80	7.43224
17	1.57935	49	4.55225	81	7.52515
18	1.67225	50	4.64515	82	7.61805
19	1.76516	51	4.73806	83	7.71095
20	1.85806	52	4.83096	84	7.80386
21	1.95096	53	4.92386	85	7.89676
22	2.04387	54	5.01676	86	7.98966
23	2.13677	55	5.10967	87	8.08256
24	2.22967	56	5.20257	88	8.17547
25	2.32258	57	5.29547	89	8.26837
26	2.41548	58	5.38838	90	8.36127
27	2.50838	59	5.48128	91	8.45418
28	2.60129	60	5.57418	92	8.54708
29	2.69419	61	5.66709	93	8.63998
30	2.78709	62	5.75999	94	8.73289
31	2.87999	63	5.85289	95	8.82579
32	2.97290	64	5.94579	96	8.91869

Table 9 *(continued)*

ft.²	m.²	ft.²	m.²	ft.²	m.²
97	9.01159	131	12.17030	165	15.32900
98	9.10450	132	12.26320	166	15.42190
99	9.19740	133	12.35610	167	15.51481
100	9.29030	134	12.44901	168	15.60771
101	9.38321	135	12.54191	169	15.70061
102	9.47611	136	12.63481	170	15.79352
103	9.56901	137	12.72772	171	15.88642
104	9.66192	138	12.82062	172	15.97932
105	9.75482	139	12.91352	173	16.07223
106	9.84772	140	13.00643	174	16.16513
107	9.94063	141	13.09933	175	16.25803
108	10.03353	142	13.19223	176	16.35093
109	10.12643	143	13.28513	177	16.44384
110	10.21933	144	13.37804	178	16.53674
111	10.31224	145	13.47094	179	16.62964
112	10.40514	146	13.56384	180	16.72255
113	10.49804	147	13.65675	181	16.81545
114	10.59095	148	13.74965	182	16.90835
115	10.68385	149	13.84255	183	17.00126
116	10.77675	150	13.93546	184	17.09416
117	10.86966	151	14.02836	185	17.18706
118	10.96256	152	14.12126	186	17.27997
119	11.05546	153	14.21417	187	17.37287
120	11.14836	154	14.30707	188	17.46577
121	11.24127	155	14.39997	189	17.55867
122	11.33417	156	14.49287	190	17.65158
123	11.42707	157	14.58578	191	17.74448
124	11.51998	158	14.67868	192	17.83738
125	11.61288	159	14.77158	193	17.93029
126	11.70578	160	14.86449	194	18.02319
127	11.79869	161	14.95739	195	18.11609
128	11.89159	162	15.05029	196	18.20900
129	11.98449	163	15.14320	197	18.30190
130	12.07740	164	15.23610	198	18.39480

Table 9 *(continued)*

ft.^{2}	m.^{2}	ft.^{2}	m.^{2}	ft.^{2}	m.^{2}
199	18.48770	233	21.64641	267	24.80511
200	18.58061	234	21.73931	268	24.89801
201	18.67351	235	21.83221	269	24.99092
202	18.76641	236	21.92512	270	25.08382
203	18.85932	237	22.01802	271	25.17672
204	18.95222	238	22.11092	272	25.26963
205	19.04512	239	22.20383	273	25.36253
206	19.13803	240	22.29673	274	25.45543
207	19.23093	241	22.38963	275	25.54834
208	19.32383	242	22.48354	276	25.64124
209	19.41674	243	22.57544	277	25.73414
210	19.50964	244	22.66834	278	25.82705
211	19.60254	245	22.76124	279	25.91995
212	19.69544	246	22.85415	280	26.01285
213	19.78835	247	22.94705	281	26.10575
214	19.88125	248	23.03995	282	26.19866
215	19.97415	249	23.13286	283	26.29156
216	20.06706	250	23.22576	284	26.38446
217	20.15996	251	23.31866	285	26.47737
218	20.25286	252	23.41157	286	26.57027
219	20.34577	253	23.50447	287	26.66317
220	20.43867	254	23.59737	288	26.75608
221	20.53157	255	23.69028	289	26.84898
222	20.62447	256	23.78318	290	26.94188
223	20.71738	257	23.87608	291	27.03478
224	20.81028	258	23.96898	292	27.12769
225	20.90318	259	24.06189	293	27.22059
226	20.99609	260	24.15479	294	27.30349
227	21.08899	261	24.24769	295	27.40640
228	21.18189	262	24.34060	296	27.49930
229	21.27480	263	24.43350	297	27.59220
230	21.36770	264	24.52640	298	27.68511
231	21.46060	265	24.61931	299	27.77801
232	21.55351	266	24.71221	300	27.87091

Table 9 *(continued)*

ft.^{2}	m.^{2}	ft.^{2}	m.^{2}	ft.^{2}	m.^{2}
301	27.96381	335	31.12252	369	34.28122
302	28.05672	336	31.21542	370	34.37412
303	28.14962	337	31.30832	371	34.46703
304	28.24252	338	31.40123	372	34.55993
305	28.33543	339	31.49413	373	34.65283
306	28.42833	340	31.58703	374	34.74574
307	28.52123	341	31.67994	375	34.83864
308	28.61414	342	31.77284	376	34.93154
309	28.70704	343	31.86574	377	35.02445
310	28.79994	344	31.95865	378	35.11735
311	28.89285	345	32.05155	379	35.21025
312	28.98575	346	32.14445	380	35.30315
313	29.07865	347	32.23735	381	35.39606
314	29.17155	348	32.33026	382	35.48896
315	29.26446	349	32.42316	383	35.58186
316	29.35736	350	32.51606	384	35.67477
317	29.45026	351	32.60897	385	35.76767
318	29.54317	352	32.70187	386	35.86057
319	29.63607	353	32.79477	387	35.85348
320	29.72897	354	32.88768	388	36.04638
321	29.82188	355	32.98058	389	36.13928
322	29.91478	356	33.07348	390	36.23219
323	30.00768	357	33.16639	391	36.32509
324	30.10058	358	33.25929	392	36.41799
325	30.19349	359	33.35219	393	36.51089
326	30.28639	360	33.44509	394	36.60380
327	30.37929	361	33.53800	395	36.69670
328	30.47220	362	33.63090	396	36.78960
329	30.56510	363	33.72380	397	36.88251
330	30.65800	364	33.81671	398	36.97541
331	30.75091	365	33.90961	399	37.06831
332	30.84381	366	34.00251	400	37.16122
333	30.93671	367	34.09542	401	37.25412
334	31.02962	368	34.18832	402	37.34702

Table 9 *(continued)*

ft.^{2}	m.^{2}	ft.^{2}	m.^{2}	ft.^{2}	m.^{2}
403	37.43993	437	40.59863	471	43.75733
404	37.53283	438	40.69153	472	43.85023
405	37.62573	439	40.78443	473	43.94314
406	37.71863	440	40.87734	474	44.03604
407	37.81154	441	40.97024	475	44.12894
408	37.90444	442	41.06314	476	44.22185
409	37.99734	443	41.15605	477	44.31475
410	38.09025	444	41.24895	478	44.40765
411	38.18315	445	41.34185	479	44.50056
412	38.27605	446	41.43476	480	44.59346
413	38.36896	447	41.52766	481	44.68636
414	38.46186	448	41.62056	482	44.77927
415	38.55476	449	41.71346	483	44.87217
416	38.64766	450	41.80637	484	44.96507
417	38.74057	451	41.89927	485	45.05797
418	38.83347	452	41.99217	486	45.15088
419	38.92637	453	42.08508	487	45.24378
420	39.01928	454	42.17798	488	45.33668
421	39.11218	455	42.27088	489	45.42959
422	39.20508	456	42.36379	490	45.52249
423	39.29799	457	42.45669	491	45.61539
424	39.39089	458	42.54959	492	45.70830
425	39.48379	459	42.64250	493	45.80120
426	39.57669	460	42.73540	494	45.89410
427	39.66960	461	42.82830	495	45.98700
428	39.76250	462	42.92120	496	46.07991
429	39.85540	463	43.01411	497	46.17281
430	39.94831	464	43.10701	498	46.26571
431	40.04121	465	43.19991	499	46.35862
432	40.13411	466	43.29282	500	46.45152
433	40.22702	467	43.38572	510	47.38055
434	40.31992	468	43.47862	520	48.30958
435	40.41282	469	43.57153	530	49.23861
436	40.50573	470	43.66443	540	50.16764

Table 9 *(continued)*

ft.^{2}	m.^{2}	ft.^{2}	m.^{2}	ft.^{2}	m.^{2}
550	51.09667	710	65.96116	860	79.89661
560	52.02570	720	66.89019	870	80.82564
570	52.95473	730	67.81922	880	81.75467
580	53.88376	740	68.74825	890	82.68371
590	54.81279	750	69.67728	900	83.61274
600	55.74182	760	70.60631	910	84.54177
610	56.67085	770	71.53534	920	85.47080
620	57.59988	780	72.46437	930	86.39983
630	58.52892	790	73.39340	940	87.32886
640	59.45795	800	74.32243	950	88.25789
650	60.38698	810	75.25146	960	89.18692
660	61.31601	820	76.18049	970	90.11595
670	62.24504	830	77.10952	980	91.04498
680	63.17407	840	78.03855	990	91.97401
690	64.10310	850	78.96758	1000	92.9030
700	65.03213				

Table 10 Square Yards to Square Meters

1 square yard = .8361 square meter

yd.^{2}	m.^{2}	yd.^{2}	m.^{2}	yd.^{2}	m.^{2}
1	0.83613	33	27.59220	65	54.34828
2	1.67225	34	28.42833	66	55.18441
3	2.50838	35	29.26446	67	56.02053
4	3.34451	36	30.10058	68	56.85666
5	4.18064	37	30.93671	69	57.69279
6	5.01676	38	31.77284	70	58.52892
7	5.85289	39	32.60897	71	59.36504
8	6.68902	40	33.44509	72	60.20117
9	7.52515	41	34.28122	73	61.03730
10	8.36127	42	35.11735	74	61.87342
11	9.19740	43	35.95348	75	62.70955
12	10.03353	44	36.78960	76	63.54568
13	10.86966	45	37.62573	77	64.38181
14	11.70578	46	38.46186	78	65.21793
15	12.54191	47	39.29799	79	66.05406
16	13.37804	48	40.13411	80	66.89019
17	14.21417	49	40.97024	81	67.72632
18	15.05029	50	41.80637	82	68.56244
19	15.88642	51	42.64250	83	69.39857
20	16.72255	52	43.47862	84	70.23470
21	17.55867	53	44.31475	85	71.07083
22	18.39480	54	45.15088	86	71.90695
23	19.23093	55	45.98700	87	72.74308
24	20.06706	56	46.82313	88	73.57921
25	20.90318	57	47.65926	89	74.41533
26	21.73931	58	48.49539	90	75.25146
27	22.57544	59	49.33151	91	76.08759
28	23.41157	60	50.16764	92	76.92372
29	24.24769	61	51.00377	93	77.75984
30	25.08382	62	51.83990	94	78.59597
31	25.91995	63	52.67602	95	79.43210
32	26.75608	64	53.51215	96	80.26823

Table 10 *(continued)*

yd.^{2}	m.^{2}	yd.^{2}	m.^{2}	yd.^{2}	m.^{2}
97	81.10435	131	109.53268	165	137.96101
98	81.94048	132	110.36881	166	138.79714
99	82.77661	133	111.20494	167	139.63327
100	83.61274	134	112.04107	168	140.46940
101	84.44886	135	112.87719	169	141.30552
102	85.28499	136	113.71332	170	142.14165
103	86.12112	137	114.54945	171	142.97778
104	86.95724	138	115.38558	172	143.81391
105	87.79337	139	116.22170	173	144.65003
106	88.62950	140	117.05783	174	145.48616
107	89.46563	141	117.89396	175	146.32229
108	90.30175	142	118.73009	176	147.15841
109	91.13788	143	119.56621	177	147.99454
110	91.97401	144	120.40234	178	148.83067
111	92.81014	145	121.23847	179	149.66680
112	93.64626	146	122.07459	180	150.50292
113	94.48239	147	122.91072	181	151.33905
114	95.31852	148	123.74685	182	152.17518
115	96.15465	149	124.58298	183	153.01131
116	96.99077	150	125.41910	184	153.84743
117	97.82690	151	126.25523	185	154.68356
118	98.66303	152	127.09136	186	155.51969
119	99.49916	153	127.92749	187	156.35582
120	100.33528	154	128.76361	188	157.19194
121	101.17141	155	129.59974	189	158.02807
122	102.00754	156	130.43587	190	158.86420
123	102.84367	157	131.27200	191	159.70033
124	103.67979	158	132.10812	192	160.53645
125	104.51592	159	132.94425	193	161.37258
126	105.35205	160	133.78038	194	162.20871
127	106.18817	161	134.61650	195	163.04483
128	107.02430	162	135.45263	196	163.88096
129	107.86043	163	136.28876	197	164.71709
130	108.69656	164	137.12489	198	165.55322

Table 10 *(continued)*

yd.^{2}	m.^{2}	yd.^{2}	m.^{2}	yd.^{2}	m.^{2}
199	166.38935	233	194.81767	267	223.24600
200	167.22547	234	195.65380	268	224.08213
201	168.06160	235	196.48993	269	224.91826
202	168.89773	236	197.32606	270	225.75439
203	169.73385	237	198.16218	271	226.59052
204	170.56998	238	198.99831	272	227.42664
205	171.40611	239	199.83444	273	228.26277
206	172.24224	240	200.67057	274	229.09890
207	173.07836	241	201.50669	275	229.93502
208	173.91449	242	202.34282	276	230.77115
209	174.75062	243	203.17895	277	231.60728
210	175.58675	244	204.01508	278	232.44341
211	176.42287	245	204.85120	279	233.27953
212	177.25900	246	205.68733	280	234.11566
213	178.09513	247	206.52346	281	234.95179
214	178.93126	248	207.35958	282	235.78792
215	179.76738	249	208.19571	283	236.62404
216	180.60351	250	209.03184	284	237.46017
217	181.43964	251	209.86797	285	238.29630
218	182.27576	252	210.70409	286	239.13243
219	183.11189	253	211.54022	287	239.96855
220	183.94802	254	212.37635	288	240.80468
221	184.78415	255	213.21248	289	241.64081
222	185.62027	256	214.04860	290	242.47693
223	186.45640	257	214.88473	291	243.31306
224	187.29253	258	215.72086	292	244.14919
225	188.12866	259	216.55699	293	244.98532
226	188.96478	260	217.39311	294	245.82144
227	189.80091	261	218.22924	295	246.65757
228	190.63704	262	219.06537	296	247.49370
229	191.47317	263	219.90149	297	248.32983
230	192.30929	264	220.73762	298	249.16595
231	193.14542	265	221.57375	299	250.00208
232	193.98155	266	222.40988	300	250.83821

Table 10 *(continued)*

yd.²	m.²	yd.²	m.²	yd.²	m.²
301	251.67434	335	280.10266	369	308.53099
302	252.51046	336	280.93879	370	309.36712
303	253.34659	337	281.77492	371	310.20325
304	254.18272	338	282.61105	372	311.03938
305	255.01884	339	283.44717	373	311.87550
306	255.85497	340	284.28330	374	312.71163
307	256.69110	341	285.11943	375	313.54776
308	257.52723	342	285.95555	376	314.38389
309	258.36335	343	286.79168	377	315.22001
310	259.19948	344	287.62781	378	316.05614
311	260.03561	345	288.46397	379	316.89227
312	260.87173	346	289.30006	380	317.72840
313	261.70786	347	290.13619	381	318.56452
314	262.54399	348	290.97232	382	319.40065
315	263.38012	349	291.80845	383	320.23678
316	264.21624	350	292.64458	384	321.07291
317	265.05237	351	293.47969	385	321.90903
318	265.88850	352	294.31683	386	322.74516
319	266.72463	353	295.15296	387	323.58129
320	267.56075	354	295.98909	388	324.41742
321	268.39688	355	296.82521	389	325.25354
322	269.23301	356	297.66134	390	326.08967
323	270.06914	357	298.49747	391	326.92580
324	270.90526	358	299.33360	392	327.76192
325	271.74139	359	300.16972	393	328.59805
326	272.57752	360	301.00585	394	329.43418
327	273.41365	361	301.84198	395	330.27031
328	274.24977	362	302.67810	396	331.10643
329	275.08590	363	303.51423	397	331.94256
330	275.92203	364	304.35036	398	332.77869
331	276.75816	365	305.18649	399	333.61481
332	277.59428	366	306.02261	400	334.45094
333	278.43041	367	306.85874	401	335.28707
334	279.26654	368	307.69487	402	336.12320

Table 10 *(continued)*

yd.²	m.²	yd.²	m.²	yd.²	m.²
403	336.95932	437	365.38766	471	393.81599
404	337.79545	438	366.22378	472	394.65211
405	338.63148	439	367.05991	473	395.48824
406	339.46771	440	367.89604	474	396.32437
407	340.30383	441	368.73217	475	397.16050
408	341.13996	442	369.56829	476	397.99662
409	341.97609	443	370.40442	477	398.83275
410	342.81222	444	371.24055	478	399.66888
411	343.64835	445	372.07668	479	400.50500
412	344.48447	446	372.91280	480	401.34113
413	345.32060	447	373.74893	481	402.17726
414	346.15673	448	374.58506	482	403.01339
415	346.99286	449	375.42118	483	403.84951
416	347.82898	450	376.25731	484	404.68564
417	348.66511	451	377.09344	485	405.52177
418	349.50124	452	377.92957	486	406.35789
419	350.33786	453	378.76569	487	407.19402
420	351.17349	454	379.60182	488	408.03015
421	352.00962	455	380.43795	489	408.86628
422	352.84575	456	381.27407	490	409.70241
423	353.68187	457	382.11020	491	410.53853
424	354.51800	458	382.94633	492	411.37466
425	355.35413	459	383.78246	493	412.21079
426	356.19025	460	384.61858	494	413.04692
427	357.02638	461	385.45471	495	413.88304
428	357.86251	462	386.29084	496	414.71917
429	358.69864	463	387.12697	497	415.55530
430	359.53476	464	387.96309	498	416.39143
431	360.37089	465	388.79922	499	417.22755
432	361.20702	466	389.63535	500	418.06368
433	362.04315	467	390.47148	510	426.42495
434	362.87927	468	391.30761	520	434.78623
435	363.71540	469	392.14373	530	443.14750
436	364.55153	470	392.97986	540	451.50877

Table 10 *(continued)*

yd.^{2}	m.^{2}	yd.^{2}	m.^{2}	yd.^{2}	m.^{2}
550	459.87005	710	593.65042	860	719.06953
560	468.23132	720	602.01170	870	727.43080
570	476.59259	730	610.37297	880	735.79208
580	484.95387	740	618.73425	890	744.15335
590	493.31514	750	627.09552	900	752.51463
600	501.67641	760	635.45679	910	760.87590
610	510.03769	770	643.81807	920	769.23717
620	518.39896	780	652.17934	930	777.59844
630	526.76024	790	660.54061	940	784.95972
640	535.12151	800	668.90189	950	794.32099
650	543.48278	810	677.26316	960	802.68227
660	551.84406	820	685.62444	970	811.04354
670	560.20533	830	693.98571	980	819.40482
680	568.56660	840	702.34698	990	827.76608
690	576.92788	850	710.70825	1000	836.12740
700	585.28915				

Table 11 Square Miles to Square Kilometers

1 square mile = 2.59 square kilometers

mi.^{2}	km.^{2}	mi.^{2}	km.^{2}	mi.^{2}	km.^{2}
1	2.5900	33	85.4696	65	168.3492
2	5.1800	34	88.0596	66	170.9392
3	7.7700	35	90.6496	67	173.5292
4	10.3600	36	93.2396	68	176.1192
5	12.9499	37	95.8296	69	178.7092
6	15.5399	38	98.4195	70	181.2992
7	18.1299	39	101.0095	71	183.8892
8	20.7199	40	103.5995	72	186.4791
9	23.3099	41	106.1895	73	189.0691
10	27.1949	42	108.7795	74	191.6591
11	28.4899	43	111.3695	75	194.2491
12	31.0799	44	113.9595	76	196.8391
13	33.6698	45	116.5495	77	199.4291
14	36.2598	46	119.1395	78	202.0191
15	38.8498	47	121.7294	79	204.6091
16	41.4398	48	124.3194	80	207.1990
17	44.0298	49	126.9094	81	209.7890
18	46.6198	50	129.4994	82	212.3790
19	49.2098	51	132.0894	83	214.9690
20	51.7998	52	134.6794	84	217.5590
21	54.3898	53	137.2694	85	220.1490
22	56.9797	54	139.8594	86	222.7390
23	59.5697	55	142.4493	87	225.3290
24	62.1597	56	145.0393	88	227.9190
25	64.7497	57	147.6293	89	230.5089
26	67.3397	58	150.2193	90	233.0989
27	69.9297	59	152.8093	91	235.6889
28	72.5197	60	155.3993	92	238.2789
29	75.1097	61	157.9893	93	240.8689
30	77.6996	62	160.5793	94	243.4589
31	80.2896	63	163.1693	95	246.0489
32	82.8796	64	165.7592	96	248.6389

Table 11 *(continued)*

mi.^{2}	km.^{2}	mi.^{2}	km.^{2}	mi.^{2}	km.^{2}
97	251.2288	132	341.8684	167	432.5280
98	253.8188	133	344.4684	168	435.1180
99	256.4088	134	347.0584	169	437.7080
100	258.9988	135	349.6484	170	440.2980
101	261.5888	136	352.2384	171	442.8880
102	264.1788	137	354.8284	172	445.4780
103	266.7688	138	357.4184	173	448.0679
104	269.3588	139	360.0083	174	450.6579
105	271.9488	140	362.5983	175	453.2479
106	274.5387	141	365.1883	176	455.8379
107	277.1287	142	367.7783	177	458.4279
108	279.7187	143	370.3683	178	461.0179
109	282.3087	144	372.9583	179	463.6079
110	284.8987	145	375.5483	180	466.1979
111	287.4887	146	378.1383	181	468.7878
112	290.0787	147	380.7283	182	471.3778
113	292.6687	148	383.3182	183	473.9678
114	295.2586	149	385.9082	184	476.5578
115	297.8486	150	388.4982	185	479.1478
116	300.4386	151	391.0882	186	481.7378
117	303.0286	152	393.6782	187	484.3278
118	305.6186	153	396.2682	188	486.9178
119	308.2086	154	398.8582	189	489.5078
120	310.7986	155	401.4482	190	492.0977
121	313.3886	156	404.0381	191	494.6877
122	315.9785	157	406.6281	192	497.2777
123	318.5685	158	409.2181	193	499.8677
124	321.1585	159	411.8081	194	502.4577
125	323.7485	160	414.3981	195	505.0477
126	326.3385	161	416.9881	196	507.6377
127	328.9285	162	419.5781	197	510.2277
128	331.5185	163	422.1681	198	512.8176
129	334.1085	164	424.7580	199	515.4076
130	336.6985	165	427.3480	200	517.9976
131	339.2884	166	429.9380		

Table 12 Acres to Hectares

1 acre = .40469 hectare or 4046.8564 square meters

acres	ha.	acres	ha.	acres	ha.
1/8	.0506	26	10.52183	58	23.47177
1/4	.1012	27	10.92651	59	23.87645
3/8	.1518	28	11.33120	60	24.28114
1/2	.2023	29	11.73588	61	24.68582
5/8	.2529	30	12.14057	62	25.09051
3/4	.3035	31	12.54525	63	25.49520
7/8	.3541	32	12.94994	64	25.89988
1	0.40469	33	13.35463	65	26.30457
2	0.80937	34	13.75931	66	26.70925
3	1.21406	35	14.16400	67	27.11394
4	1.61874	36	14.56868	68	27.51862
5	2.02343	37	14.97337	69	27.92331
6	2.42811	38	15.37805	70	28.32799
7	2.83280	39	15.78274	71	28.73268
8	3.23749	40	16.18743	72	29.13737
9	3.64217	41	16.59211	73	29.54205
10	4.04686	42	16.99680	74	29.94674
11	4.45154	43	17.40148	75	30.35142
12	4.85623	44	17.80617	76	30.75611
13	5.26091	45	18.21085	77	31.16079
14	5.66560	46	18.61554	78	31.56548
15	6.07028	47	19.02023	79	31.97017
16	6.47497	48	19.42491	80	32.37485
17	6.87966	49	19.82960	81	32.77954
18	7.28434	50	20.23428	82	33.18422
19	7.68903	51	20.63897	83	33.58891
20	8.09371	52	21.04365	84	33.99359
21	8.49840	53	21.44834	85	34.39828
22	8.90308	54	21.85302	86	34.80297
23	9.30777	55	22.25771	87	35.20765
24	9.71246	56	22.66240	88	35.61234
25	10.11714	57	23.06708	89	36.01702

Table 12 *(continued)*

acres	ha.	acres	ha.	acres	ha.
90	36.42171	210	84.98398	410	165.92111
91	36.82639	220	89.03084	420	169.96797
92	37.23108	230	93.07770	430	174.01483
93	37.63576	240	97.12455	440	178.06168
94	38.04045	250	101.17141	450	182.10854
95	38.44514	260	105.21827	460	186.15540
96	38.84982	270	109.26512	470	190.20225
97	39.25451	280	113.31198	480	194.24911
98	39.65919	290	117.35884	490	198.29596
99	40.06388	300	121.40569	500	202.34282
100	40.46856	310	125.45255	550	222.57710
110	44.51542	320	129.49940	600	242.81138
120	48.56228	330	133.54626	650	263.04567
130	52.60913	340	137.59312	700	283.27995
140	56.65599	350	141.63997	750	303.51423
150	60.70285	360	145.68683	800	323.74851
160	64.74970	370	149.73369	850	343.98280
170	68.79656	380	153.78054	900	364.21708
180	72.84342	390	157.82740	950	384.45136
190	76.89027	400	161.87426	1000	404.6856
200	80.93713				

Weight

Table 13 Grains to Milligrams

1 grain = 64.80 milligram

grains	mg.	grains	mg.	grains	mg.
1/4	16.20	30	1943.97	62	4017.53
1/2	32.50	31	2008.77	63	4082.33
3/4	48.60	32	2073.57	64	4147.13
1	64.80	33	2138.36	65	4211.93
2	129.60	34	2203.16	66	4276.73
3	194.40	35	2267.96	67	4341.53
4	259.20	36	2332.76	68	4406.33
5	323.99	37	2397.56	69	4471.12
6	388.79	38	2462.36	70	4535.92
7	453.59	39	2527.16	71	4600.72
8	518.39	40	2591.96	72	4665.52
9	583.19	41	2656.76	73	4730.32
10	647.99	42	2721.55	74	4795.12
11	712.79	43	2786.35	75	4859.92
12	777.59	44	2851.15	76	4924.72
13	842.39	45	2915.95	77	4989.52
14	907.18	46	2980.75	78	5054.31
15	971.98	47	3045.55	79	5119.11
16	1036.78	48	3110.35	80	5183.91
17	1101.58	49	3175.15	81	5248.71
18	1166.38	50	3239.95	82	5313.51
19	1231.18	51	3304.74	83	5378.31
20	1295.98	52	3369.54	84	5443.11
21	1360.78	53	3434.34	85	5507.91
22	1425.58	54	3499.14	86	5572.71
23	1490.37	55	3563.94	87	5637.51
24	1555.17	56	3628.74	88	5702.30
25	1619.97	57	3693.54	89	5767.10
26	1684.77	58	3758.34	90	5831.90
27	1749.57	59	3823.14	91	5896.70
28	1814.37	60	3887.93	92	5961.50
29	1879.17	61	3952.73	93	6026.30

Table 13 *(continued)*

grains	mg.	grains	mg.	grains	mg.
94	6091.10	128	8294.26	162	10497.42
95	6155.90	129	8359.06	163	10562.22
96	6220.70	130	8423.86	164	10627.02
97	6285.49	131	8488.66	165	10691.82
98	6350.29	132	8553.46	166	10756.62
99	6415.09	133	8618.26	167	10821.42
100	6479.89	134	8683.05	168	10886.22
101	6544.69	135	8747.85	169	10951.02
102	6609.49	136	8812.65	170	11015.81
103	6674.29	137	8877.45	171	11080.61
104	6739.09	138	8942.25	172	11145.41
105	6803.89	139	9007.05	173	11210.21
106	6868.68	140	9071.85	174	11275.01
107	6933.48	141	9136.65	175	11339.81
108	6998.28	142	9201.45	176	11404.61
109	7063.08	143	9266.24	177	11469.41
110	7127.88	144	9331.04	178	11534.21
111	7192.68	145	9395.85	179	11599.00
112	7257.48	146	9460.64	180	11663.80
113	7322.28	147	9525.44	181	11728.60
114	7387.08	148	9590.24	182	11793.40
115	7451.87	149	9655.04	183	11858.20
116	7516.67	150	9719.84	184	11923.00
117	7581.47	151	9784.64	185	11987.80
118	7646.27	152	9849.43	186	12052.60
119	7711.07	153	9914.24	187	12117.40
120	7775.87	154	9979.03	188	12182.20
121	7840.67	155	10043.83	189	12246.99
122	7905.47	156	10108.63	190	12311.79
123	7970.27	157	10173.43	191	12376.59
124	8035.06	158	10238.23	192	12441.39
125	8099.86	159	10303.03	193	12506.19
126	8164.66	160	10367.83	194	12570.99
127	8229.46	161	10432.62	195	12635.79

Table 13 *(continued)*

grains	mg.	grains	mg.	grains	mg.
196	12700.59	230	14903.75	264	17106.91
197	12765.39	231	14968.55	265	17171.71
198	12830.18	232	15033.35	266	17236.51
199	12894.98	233	15098.15	267	17301.31
200	12959.78	234	15162.94	268	17366.11
201	13024.58	235	15227.74	269	17430.91
202	13089.38	236	15292.54	270	17495.71
203	13154.18	237	15357.34	271	17560.50
204	13218.98	238	15422.14	272	17625.30
205	13283.78	239	15486.94	273	17690.10
206	13348.58	240	15551.74	274	17754.90
207	13413.37	241	15616.54	275	17819.70
208	13478.17	242	15681.34	276	17884.50
209	13542.97	243	15746.14	277	17949.30
210	13607.77	244	15810.93	278	18014.10
211	13672.57	245	15875.73	279	18078.90
212	13737.37	246	15940.53	280	18143.69
213	13802.17	247	16005.33	281	18208.49
214	13866.97	248	16070.13	282	18273.29
215	13931.77	249	16134.93	283	18338.09
216	13996.56	250	16199.73	284	18402.89
217	14061.36	251	16264.53	285	18467.69
218	14126.16	252	16329.33	286	18532.49
219	14190.96	253	16394.12	287	18597.29
220	14255.76	254	16458.92	288	18662.09
221	14320.56	255	16523.72	289	18726.89
222	14385.36	256	16588.52	290	18791.68
223	14450.16	257	16653.32	291	18856.48
224	14514.96	258	16718.12	292	18921.28
225	14579.75	259	16782.92	293	18986.08
226	14644.55	260	16847.72	294	19050.88
227	14709.35	261	16912.52	295	19115.68
228	14774.15	262	16977.31	296	19180.48
229	14838.95	263	17042.11	297	19245.28

Table 13 *(continued)*

grains	mg.	grains	mg.	grains	mg.
298	19310.08	332	21513.24	366	23716.40
299	19374.87	333	21578.04	367	23781.20
300	19439.67	334	21642.84	368	23846.00
301	19504.47	335	21707.63	369	23910.80
302	19569.27	336	21772.43	370	23975.60
303	19634.07	337	21837.23	371	24040.40
304	19698.87	338	21902.03	372	24105.19
305	19763.67	339	21966.83	373	24169.99
306	19828.47	340	22031.63	374	24234.79
307	19893.27	341	22096.43	375	24299.59
308	19958.06	342	22161.23	376	24364.39
309	20022.86	343	22226.03	377	24429.19
310	20087.66	344	22290.82	378	24493.99
311	20152.46	345	22355.62	379	24558.79
312	20217.26	346	22420.42	380	24623.59
313	20282.06	347	22485.22	381	24688.38
314	20346.86	348	22550.02	382	24753.18
315	20411.66	349	22614.82	383	24817.98
316	20476.46	350	22679.62	384	24882.78
317	20541.25	351	22744.42	385	24947.58
318	20606.05	352	22809.22	386	25012.38
319	20670.84	353	22874.02	387	25077.18
320	20735.65	354	22938.81	388	25141.98
321	20800.45	355	23003.61	389	25206.78
322	20865.25	356	23068.41	390	25271.57
323	20930.05	357	23133.21	391	25336.37
324	20994.85	358	23198.01	392	25401.17
325	21059.65	359	23262.81	393	25465.97
326	21124.44	360	23327.61	394	25530.77
327	21189.24	361	23392.41	395	25595.57
328	21254.04	362	23457.21	396	25660.37
329	21318.84	363	23522.00	397	25725.17
330	21383.64	364	23586.80	398	25789.97
331	21448.44	365	23651.60	399	25854.77

Table 13 *(continued)*

grains	mg.	grains	mg.	grains	mg.
400	25919.56	434	28122.73	468	30325.89
401	25984.36	435	28187.53	469	30390.69
402	26049.16	436	28252.32	470	30455.49
403	26113.96	437	28317.12	471	30520.29
404	26178.76	438	28381.92	472	30585.09
405	26243.56	439	28446.72	473	30649.88
406	26308.36	440	28511.52	474	30714.68
407	26373.16	441	28576.32	475	30779.48
408	26437.96	442	28641.12	476	30844.28
409	26502.75	443	28705.92	477	30909.08
410	26567.55	444	28770.72	478	30973.88
411	26632.35	445	28835.51	479	31038.68
412	26697.15	446	28900.31	480	31103.48
413	26761.95	447	28965.11	481	31168.28
414	26826.75	448	29029.91	482	31233.07
415	26891.55	449	29094.71	483	31297.87
416	26956.35	450	29159.51	484	31362.67
417	27021.15	451	29224.31	485	31427.47
418	27085.94	452	29289.11	486	31492.27
419	27150.74	453	29353.91	487	31557.07
420	27215.54	454	29418.71	488	31621.87
421	27280.34	455	29483.50	489	31686.67
422	27345.14	456	29548.30	490	31751.47
423	27409.94	457	29613.10	491	31816.26
424	27474.74	458	29677.90	492	31881.06
425	27539.54	459	29742.70	493	31945.86
426	27604.34	460	29807.50	494	32010.66
427	27669.13	461	29872.30	495	32075.46
428	27733.93	462	29937.10	496	32140.26
429	27798.73	463	30001.90	497	32205.06
430	27863.53	464	30066.69	498	32269.86
431	27928.33	465	30131.49	499	32334.66
432	27993.13	466	30196.29	500	32399.45
433	28057.93	467	30261.09	510	33047.44

Table 13 *(continued)*

grains	mg.	grains	mg.	grains	mg.
520	33695.43	690	44711.25	850	55079.07
530	34343.42	700	45359.24	860	55727.06
540	34991.41	710	46007.23	870	56375.05
550	35639.40	720	46655.21	880	57023.04
560	36287.39	730	47303.20	890	57671.03
570	36935.38	740	47951.19	900	58319.02
580	37583.37	750	48599.18	910	58967.01
590	38231.36	760	49247.17	920	59615.00
600	38879.35	770	49895.16	930	60262.99
610	39527.33	780	50543.15	940	60910.98
620	40175.32	790	51191.14	950	61558.96
630	40823.31	800	51839.13	960	62206.95
640	41471.30	810	52487.12	970	62854.94
650	42119.29	820	53135.11	980	63502.93
660	42767.28	830	53783.10	990	64150.92
670	43415.27	840	54431.08	1000	64798.90
680	44063.26				

Table 14 Avoirdupois Pounds, Ounces to Grams

Note: For a comparison among weight values of avoirdupois and troy pounds and ounces, please see "Weight Conversion Tables, Customary to Customary," in Part III.

1 avoirdupois pound = 453.592 grams
1 avoirdupois ounce = 28.350 grams

lb.	oz.	g.	lb.	oz.	g.	lb.	oz.	g.
	1/4	7.0874	1	10	737.088	3	6	1530.874
	1/2	14.1748	1	11	765.437	3	7	1559.224
	3/4	21.2621	1	12	793.787	3	8	1587.573
	1	28.350	1	13	822.136	3	9	1615.923
	2	56.699	1	14	850.486	3	10	1644.272
	3	85.049	1	15	878.835	3	11	1672.622
	4	113.398	2		907.185	3	12	1700.971
	5	141.748	2	1	935.534	3	13	1729.321
	6	170.097	2	2	963.884	3	14	1757.670
	7	198.447	2	3	992.233	3	15	1786.020
	8	226.796	2	4	1020.583	4		1814.369
	9	255.146	2	5	1048.932	4	1	1842.719
	10	283.495	2	6	1077.282	4	2	1871.069
	11	311.845	2	7	1105.631	4	3	1899.418
	12	340.194	2	8	1133.981	4	4	1927.768
	13	368.544	2	9	1162.330	4	5	1956.117
	14	396.893	2	10	1190.680	4	6	1984.467
	15	425.243	2	11	1219.029	4	7	2012.816
1		453.592	2	12	1247.379	4	8	2041.166
1	1	481.942	2	13	1275.729	4	9	2069.515
1	2	510.291	2	14	1304.078	4	10	2097.865
1	3	538.641	2	15	1332.428	4	11	2126.214
1	4	566.990	3		1360.777	4	12	2154.564
1	5	595.340	3	1	1389.127	4	13	2182.913
1	6	623.690	3	2	1417.476	4	14	2211.263
1	7	652.039	3	3	1445.826	4	15	2239.612
1	8	680.389	3	4	1474.175	5		2267.962
1	9	708.738	3	5	1502.525	5	1	2296.311

Table 14 *(continued)*

lb.	oz.	g.	lb.	oz.	g.	lb.	oz.	g.
5	2	2324.661	6	13	3090.098	8	7	3827.186
5	3	2353.010	6	14	3118.448	8	8	3855.535
5	4	2381.360	6	15	3146.797	8	9	3883.885
5	5	2409.709	7		3175.147	8	10	3912.234
5	6	2438.059	7	1	3203.496	8	11	3940.584
5	7	2466.408	7	2	3231.846	8	12	3968.933
5	8	2494.758	7	3	3260.195	8	13	3997.283
5	9	2523.108	7	4	3288.545	8	14	4025.632
5	10	2551.457	7	5	3316.894	8	15	4053.982
5	11	2579.807	7	6	3345.244	9		4082.331
5	12	2608.156	7	7	3373.593	9	1	4110.681
5	13	2636.506	7	8	3401.943	9	2	4139.030
5	14	2664.855	7	9	3430.292	9	3	4167.380
5	15	2693.205	7	10	3458.642	9	4	4195.729
6		2721.554	7	11	3486.991	9	5	4224.079
6	1	2749.904	7	12	3515.341	9	6	4252.428
6	2	2778.253	7	13	3543.690	9	7	4280.778
6	3	2806.603	7	14	3572.040	9	8	4309.128
6	4	2834.952	7	15	3600.389	9	9	4337.477
6	5	2863.302	8		3628.739	9	10	4365.827
6	6	2891.651	8	1	3657.008	9	11	4394.176
6	7	2920.001	8	2	3685.438	9	12	4422.526
6	8	2948.350	8	3	3713.788	9	13	4450.875
6	9	2976.700	8	4	3742.137	9	14	4479.225
6	10	3005.049	8	5	3770.487	9	15	4507.574
6	11	3033.399	8	6	3798.836	10		4535.924
6	12	3061.748						

Table 15 Troy Ounces to Grams

Note: For a comparison among weight values of avoirdupois and troy pounds and ounces, please see "Weight Conversion Tables, Customary to Customary" in Part III.

1 troy ounce = 31.103 grams

troy oz.	g.	troy oz.	g.	troy oz.	g.
1/4	7.776	26	808.690	54	1679.588
1/2	15.552	27	839.794	55	1710.691
3/4	23.328	28	870.897	56	1741.795
1	31.103	29	902.001	57	1772.898
2	62.207	30	933.104	58	1804.002
3	93.310	31	964.208	59	1835.105
4	124.414	32	995.311	60	1866.209
5	155.517	33	1026.415	61	1897.312
6	186.621	34	1057.518	62	1928.416
7	217.724	35	1088.622	63	1959.519
8	248.828	36	1119.725	64	1990.623
9	279.931	37	1150.829	65	2021.726
10	311.035	38	1181.932	66	2052.829
11	342.138	39	1213.036	67	2083.933
12	373.242	40	1244.139	68	2115.036
13	404.345	41	1275.243	69	2146.140
14	435.449	42	1306.346	70	2177.243
15	466.552	43	1337.449	71	2208.347
16	497.656	44	1368.553	72	2239.450
17	528.759	45	1399.656	73	2270.554
18	559.863	46	1430.760	74	2301.657
19	590.966	47	1461.863	75	2332.761
20	622.070	48	1492.967	76	2363.864
21	653.173	49	1524.070	77	2394.968
22	684.276	50	1555.174	78	2426.071
23	715.380	51	1586.277	79	2457.175
24	746.483	52	1617.381	80	2488.278
25	777.587	53	1648.484	81	2519.382

Table 15 *(continued)*

troy oz.	g.	troy oz.	g.	troy oz.	g.
82	2550.485	116	3608.003	150	4665.521
83	2581.589	117	3639.107	151	4696.625
84	2612.692	118	3670.210	152	4727.728
85	2643.796	119	3701.314	153	4758.832
86	2674.899	120	3732.417	154	4789.935
87	2706.002	121	3763.521	155	4821.039
88	2737.106	122	3794.624	156	4852.142
89	2768.209	123	3825.728	157	4883.246
90	2799.313	124	3856.831	158	4914.349
91	2830.416	125	3887.935	159	4945.453
92	2861.520	126	3919.038	160	4976.556
93	2892.623	127	3950.142	161	5007.660
94	2923.727	128	3981.245	162	5038.763
95	2954.830	129	4012.348	163	5069.867
96	2985.934	130	4043.452	164	5100.970
97	3017.037	131	4074.555	165	5132.074
98	3048.141	132	4105.659	166	5163.177
99	3079.244	133	4136.762	167	5194.281
100	3110.348	134	4167.866	168	5225.384
101	3141.451	135	4198.969	169	5256.488
102	3172.555	136	4230.073	170	5287.591
103	3203.658	137	4261.176	171	5318.695
104	3234.762	138	4292.280	172	5349.798
105	3265.865	139	4323.383	173	5380.901
106	3296.969	140	4354.487	174	5412.005
107	3328.072	141	4385.590	175	5443.108
108	3359.175	142	4416.694	176	5474.212
109	3390.279	143	4447.797	177	5505.315
110	3421.382	144	4478.901	178	5536.419
111	3452.486	145	4510.004	179	5567.522
112	3483.589	146	4541.108	180	5598.626
113	3514.693	147	4572.211	181	5629.729
114	3545.796	148	4603.315	182	5660.833
115	3576.900	149	4634.418	183	5691.936

Table 15 *(continued)*

troy oz.	g.	troy oz.	g.	troy oz.	g.
184	5723.040	190	5909.661	196	6096.281
185	5754.143	191	5940.764	197	6127.385
186	5785.247	192	5971.867	198	6158.488
187	5816.350	193	6002.971	199	6189.592
188	5847.454	194	6034.074	200	6220.695
189	5878.557	195	6065.178		

Table 16 Avoirdupois Pounds to Kilograms

1 pound = .4536 kilogram

lb.	kg.	lb.	kg.	lb.	kg.
1	0.45359	33	14.96855	65	29.48350
2	0.90718	34	15.42214	66	29.93710
3	1.36078	35	15.87573	67	30.39069
4	1.81437	36	16.32933	68	30.84428
5	2.26796	37	16.78292	69	31.29787
6	2.72155	38	17.23651	70	31.75147
7	3.17515	39	17.69010	71	32.20506
8	3.62874	40	18.14369	72	32.65865
9	4.08233	41	18.59729	73	33.11224
10	4.53592	42	19.05088	74	33.56583
11	4.98952	43	19.50447	75	34.01943
12	5.44311	44	19.95806	76	34.47302
13	5.89670	45	20.41166	77	34.92661
14	6.35029	46	20.86525	78	35.38020
15	6.80389	47	21.31884	79	35.83380
16	7.25748	48	21.77243	80	36.28739
17	7.71107	49	22.22603	81	36.74098
18	8.16466	50	22.67962	82	37.19457
19	8.61826	51	23.13321	83	37.64817
20	9.07185	52	23.58680	84	38.10176
21	9.52544	53	24.04040	85	38.55535
22	9.97903	54	24.49399	86	39.00894
23	10.43262	55	24.94758	87	39.46254
24	10.88622	56	25.40117	88	39.91613
25	11.33981	57	25.85476	89	40.36972
26	11.79340	58	26.30836	90	40.82331
27	12.24699	59	26.76195	91	41.27691
28	12.70059	60	27.21554	92	41.73050
29	13.15418	61	27.66913	93	42.18409
30	13.60777	62	28.12273	94	42.63768
31	14.06136	63	28.57632	95	43.09128
32	14.51496	64	29.02991	96	43.54487

Table 16 *(continued)*

lb.	kg.	lb.	kg.	lb.	kg.
97	43.99846	131	59.42060	165	74.84274
98	44.45205	132	59.87419	166	75.29633
99	44.90564	133	60.32779	167	75.74993
100	45.35924	134	60.78138	168	76.20352
101	45.81283	135	61.23497	169	76.65711
102	46.26642	136	61.68856	170	77.11070
103	46.72001	137	62.14215	171	77.56429
104	47.17361	138	62.59575	172	78.01789
105	47.62720	139	63.04934	173	78.47148
106	48.08079	140	63.50293	174	78.92507
107	48.53438	141	63.95652	175	79.37866
108	48.98798	142	64.41012	176	79.83226
109	49.44157	143	64.86371	177	80.28585
110	49.89516	144	65.31730	178	80.73944
111	50.34875	145	65.77089	179	81.19303
112	50.80235	146	66.22449	180	81.64663
113	51.25594	147	66.67808	181	82.10022
114	51.70953	148	67.13167	182	82.55381
115	52.16312	149	67.58526	183	83.00740
116	52.61671	150	68.03886	184	83.46100
117	53.07031	151	68.49245	185	83.91459
118	53.52390	152	68.94604	186	84.36818
119	53.97749	153	69.39963	187	84.82177
120	54.43108	154	69.85322	188	85.27536
121	54.88468	155	70.30682	189	85.72896
122	55.33827	156	70.76041	190	86.18255
123	55.79186	157	71.21400	191	86.63614
124	56.24545	158	71.66759	192	87.08974
125	56.69905	159	72.12119	193	87.54333
126	57.15264	160	72.57478	194	87.99692
127	57.60623	161	73.02837	195	88.45051
128	58.05982	162	73.48196	196	88.90410
129	58.51342	163	73.93556	197	89.35770
130	58.96701	164	74.38915	198	89.81129

Table 16 *(continued)*

lb.	kg.	lb.	kg.	lb.	kg.
199	90.26488	233	105.68702	267	121.10916
200	90.71847	234	106.14061	268	121.56275
201	91.17207	235	106.59421	269	122.01635
202	91.62566	236	107.04780	270	122.46994
203	92.07925	237	107.50139	271	122.92353
204	92.53284	238	107.95498	272	123.37712
205	92.98644	239	108.40858	273	123.83072
206	93.44003	240	108.86217	274	124.28431
207	93.89362	241	109.31576	275	124.73790
208	94.34721	242	109.76935	276	125.19149
209	94.80081	243	110.22295	277	125.64509
210	95.25440	244	110.67654	278	126.09868
211	95.70799	245	111.13013	279	126.55227
212	96.16158	246	111.58372	280	127.00586
213	96.61517	247	112.03732	281	127.45946
214	97.06877	248	112.49091	282	127.91305
215	97.52236	249	112.94450	283	128.36664
216	97.97595	250	113.39809	284	128.82023
217	98.42954	251	113.85168	285	129.27382
218	98.88314	252	114.30528	286	129.72742
219	99.33673	253	114.75887	287	130.18101
220	99.79032	254	115.21246	288	130.63460
221	100.24391	255	115.66605	289	131.08819
222	100.69751	256	116.11965	290	131.54179
223	101.15110	257	116.57324	291	131.99538
224	101.60469	258	117.02683	292	132.44897
225	102.05828	259	117.48042	293	132.90256
226	102.51188	260	117.93402	294	133.35616
227	102.96547	261	118.38761	295	133.80975
228	103.41906	262	118.84120	296	134.26334
229	103.87265	263	119.29479	297	134.71693
230	104.32625	264	119.74839	298	135.17053
231	104.77984	265	120.20198	299	135.62414
232	105.23343	266	120.65557	300	136.07771

Table 16 *(continued)*

lb.	kg.	lb.	kg.	lb.	kg.
301	136.53130	335	151.95344	369	167.37558
302	136.98490	336	152.40704	370	167.82918
303	137.43849	337	152.86063	371	168.28277
304	137.89208	338	153.31422	372	168.73636
305	138.34567	339	153.76781	373	169.18995
306	138.79926	340	154.22141	374	169.64355
307	139.25286	341	154.67500	375	170.09714
308	139.70645	342	155.12859	376	170.55073
309	140.16004	343	155.58218	377	171.00432
310	140.61363	344	156.03577	378	171.45792
311	141.06723	345	156.48937	379	171.91151
312	141.52082	346	156.94296	380	172.36510
313	141.97441	347	157.39655	381	172.81869
314	142.42800	348	157.85014	382	173.27229
315	142.88160	349	158.30374	383	173.72588
316	143.33519	350	158.75733	384	174.17947
317	143.78878	351	159.21092	385	174.63306
318	144.24237	352	159.66451	386	175.08665
319	144.69596	353	160.11811	387	175.54025
320	145.14956	354	160.57170	388	175.99384
321	145.60315	355	161.02529	389	176.44743
322	146.05674	356	161.47888	390	176.90102
323	146.51033	357	161.93248	391	177.35462
324	146.96393	358	162.38607	392	177.80821
325	147.41752	359	162.83966	393	178.26180
326	147.87111	360	163.29325	394	178.71539
327	148.32471	361	163.74685	395	179.16899
328	148.77830	362	164.20044	396	179.62258
329	149.23189	363	164.65403	397	180.07617
330	149.68548	364	165.10762	398	180.52976
331	150.13907	365	165.56121	399	180.98335
332	150.59267	366	166.01481	400	181.43695
333	151.04626	367	166.46840	401	181.89054
334	151.49985	368	166.92199	402	182.34413

Table 16 *(continued)*

lb.	kg.	lb.	kg.	lb.	kg.
403	182.79772	437	198.21987	471	213.64201
404	183.25132	438	198.67346	472	214.09560
405	183.70491	439	199.12705	473	214.54919
406	184.15850	440	199.58064	474	215.00278
407	184.61209	441	200.03424	475	215.45638
408	185.06569	442	200.48783	476	215.90997
409	185.51928	443	200.94142	477	216.36356
410	185.97287	444	201.39501	478	216.81715
411	186.42646	445	201.84860	479	217.27074
412	186.88006	446	202.30220	480	217.72434
413	187.33365	447	202.75579	481	218.17793
414	187.78724	448	203.20938	482	218.63152
415	188.24083	449	203.66297	483	219.08511
416	188.69443	450	204.11657	484	219.53871
417	189.14802	451	204.57016	485	219.99230
418	189.60161	452	205.02375	486	220.44589
419	190.05520	453	205.47734	487	220.89948
420	190.50879	454	205.93093	488	221.35308
421	190.96239	455	206.38453	489	221.80667
422	191.41598	456	206.83812	490	222.26026
423	191.86957	457	207.29171	491	222.71385
424	192.32316	458	207.74530	492	223.16745
425	192.77676	459	208.19890	493	223.62104
426	193.23035	460	208.65249	494	224.07463
427	193.68394	461	209.10608	495	224.52822
428	194.13753	462	209.55968	496	224.98182
429	194.59113	463	210.01327	497	225.43541
430	195.04472	464	210.46686	498	225.88900
431	195.49831	465	210.92045	499	226.34259
432	195.95190	466	211.37404	500	226.79618
433	196.40550	467	211.82764	510	231.33211
434	196.85909	468	212.28123	520	235.86803
435	197.31268	469	212.73482	530	240.40396
436	197.76627	470	213.18841	540	244.93988

Table 16 *(continued)*

lb.	kg.	lb.	kg.	lb.	kg.
550	249.47580	710	322.05058	860	390.08944
560	254.01173	720	326.58651	870	394.62536
570	258.54765	730	331.12243	880	399.16129
580	263.08357	740	335.65835	890	403.69721
590	267.61950	750	340.19427	900	408.23313
600	272.15542	760	344.73020	910	412.76905
610	276.69135	770	349.26612	920	417.30498
620	281.22727	780	353.80205	930	421.84090
630	285.76319	790	358.33797	940	426.37683
640	290.29911	800	362.87389	950	430.91275
650	294.83504	810	367.40982	960	435.44867
660	299.37096	820	371.94574	970	439.98460
670	303.90689	830	376.48167	980	444.52052
680	308.44281	840	381.01759	990	449.05645
690	312.97873	850	385.55351	1000	453.5924
700	317.51466				

Table 17 Short Tons to Metric Tons

1 short ton = .9072 metric ton

tons	t.	tons	t.	tons	t.
½	.4536	23	20.8652	55	49.8952
1	.9072	24	21.7724	56	50.8023
1½	1.3608	25	22.6796	57	51.7095
2	1.8144	26	23.5868	58	52.6167
2½	2.2680	27	24.4940	59	53.5239
3	2.7216	28	25.4012	60	54.4311
3½	3.1751	29	26.3084	61	55.3383
4	3.6287	30	27.2155	62	56.2455
4½	4.0823	31	28.1227	63	57.1526
5	4.5359	32	29.0299	64	58.0598
5½	4.9895	33	29.9371	65	58.9670
6	5.4431	34	30.8443	66	59.8742
6½	5.8967	35	31.7515	67	60.7814
7	6.3503	36	32.6587	68	61.6886
7½	6.8039	37	33.5658	69	62.5957
8	7.2575	38	34.4730	70	63.5029
8½	7.7111	39	35.3802	71	64.4101
9	8.1647	40	36.2874	72	65.3173
9½	8.6183	41	37.1946	73	66.2245
10	9.0718	42	38.1018	74	67.1317
11	9.9790	43	39.0089	75	68.0389
12	10.8862	44	39.9161	76	68.9460
13	11.7934	45	40.8233	77	69.8532
14	12.7006	46	41.7305	78	70.7604
15	13.6078	47	42.6377	79	71.6676
16	14.5150	48	43.5449	80	72.5748
17	15.4221	49	44.4521	81	73.4820
18	16.3293	50	45.3592	82	74.3891
19	17.2365	51	46.2664	83	75.2963
20	18.1437	52	47.1736	84	76.2035
21	19.0509	53	48.0808	85	77.1107
22	19.9581	54	48.9880	86	78.0179

Table 17 *(continued)*

tons	t.	tons	t.	tons	t.
87	78.9251	121	109.7694	155	140.6136
88	79.8323	122	110.6765	156	141.5208
89	80.7394	123	111.5837	157	142.4280
90	81.6466	124	112.4909	158	143.3352
91	82.5538	125	113.3981	159	144.2424
92	83.4610	126	114.3053	160	145.1496
93	84.3682	127	115.2125	161	146.0567
94	85.2754	128	116.1196	162	146.9639
95	86.1826	129	117.0268	163	147.8711
96	87.0897	130	117.9340	164	148.7783
97	87.9969	131	118.8412	165	149.6855
98	88.9041	132	119.7484	166	150.5927
99	89.8113	133	120.6556	167	151.4999
100	90.7185	134	121.5628	168	152.4070
101	91.6257	135	122.4699	169	153.3142
102	92.5328	136	123.3771	170	154.2214
103	93.4400	137	124.2843	171	155.1286
104	94.3472	138	125.1915	172	156.0358
105	95.2544	139	126.0987	173	156.9430
106	96.1616	140	127.0059	174	157.8501
107	97.0688	141	127.9130	175	158.7573
108	97.9760	142	128.8202	176	159.6645
109	98.8831	143	129.7274	177	160.5717
110	99.7903	144	130.6346	178	161.4789
111	100.6975	145	131.5418	179	162.3861
112	101.6047	146	132.4490	180	163.2933
113	102.5119	147	133.3562	181	164.2004
114	103.4191	148	134.2633	182	165.1076
115	104.3262	149	135.1705	183	166.0148
116	105.2334	150	136.0777	184	166.9220
117	106.1406	151	136.9849	185	167.8292
118	107.0478	152	137.8921	186	168.7364
119	107.9550	153	138.7993	187	169.6435
120	108.8622	154	139.7064	188	170.5507

Table 17 *(continued)*

tons	t.	tons	t.	tons	t.
189	171.4579	199	180.5298	2000	1814.3695
190	172.3651	200	181.4369	3000	2721.5542
191	173.2723	300	272.1554	4000	3628.7390
192	174.1795	400	362.8739	5000	4535.9237
193	175.0867	500	453.5924	6000	5443.1084
194	175.9938	600	544.3108	7000	6350.2932
195	176.9010	700	635.0293	8000	7257.4779
196	177.8082	800	725.7478	9000	8164.6627
197	178.7154	900	816.4663	10000	9071.8474
198	179.6226	1000	907.1847		

Capacity

Table 18 Fluid Ounces to Cups to Milliliters

1 fluid ounce = 29.57 milliliters
1 cup = 236.59 milliliters

fl. oz.	cups	ml.	fl. oz.	cups	ml.
¼		7.39	29		857.63
½		14.79	30		887.21
¾		22.18	31		916.78
1	⅛	29.57	32	4	946.35
2	¼	59.15	33		975.93
3	⅜	88.72	34		1005.50
4	½	118.29	35		1035.07
5	⅝	147.87	36		1064.65
6	¾	177.44	37		1094.22
7	⅞	207.01	38		1123.79
8	1	236.59	39		1153.37
9		266.16	40	5	1182.94
10		295.74	41		1212.51
11		325.31	42		1242.09
12		354.88	43		1271.66
13		384.46	44		1301.24
14		414.03	45		1330.81
15		443.60	46		1360.38
16	2	473.18	47		1389.96
17		502.75	48	6	1419.53
18		532.32	49		1449.10
19		561.90	50		1478.68
20		591.47	51		1508.25
21		621.04	52		1537.82
22		650.62	53		1567.40
23		680.19	54		1596.97
24	3	709.76	55		1626.54
25		739.34	56	7	1656.12
26		768.91	57		1685.69
27		798.49	58		1715.26
28		828.06	59		1744.84

Table 18 *(continued)*

fl. oz.	cups	ml.	fl. oz.	cups	ml.
60		1774.41	63		1863.13
61		1803.99	64	8	1892.71
62		1833.56			

Table 19 Liquid Quarts to Liters

1 liquid quart = .94635 liter

qt.	l.	qt.	l.	qt.	l.
1	0.94635	33	31.22965	65	61.51294
2	1.89271	34	32.17600	66	62.45929
3	2.83906	35	33.12235	67	63.40565
4	3.78541	36	34.06871	68	64.35200
5	4.73176	37	35.01506	69	65.29835
6	5.67812	38	35.96141	70	66.24471
7	6.62447	39	36.90776	71	67.19106
8	7.57082	40	37.85412	72	68.13741
9	8.51718	41	38.80047	73	69.08376
10	9.46353	42	39.74682	74	70.03012
11	10.40988	43	40.69318	75	70.97647
12	11.35624	44	41.63953	76	71.92282
13	12.30259	45	42.58588	77	72.86918
14	13.24894	46	43.53224	78	73.81553
15	14.19529	47	44.47859	79	74.76188
16	15.14165	48	45.42494	80	75.70823
17	16.08800	49	46.37129	81	76.65469
18	17.03435	50	47.31765	82	77.60094
19	17.98071	51	48.26400	83	78.54729
20	18.92706	52	49.21035	84	79.49365
21	19.87341	53	50.15671	85	80.44000
22	20.81976	54	51.10306	86	81.38635
23	21.76612	55	52.04941	87	82.33271
24	22.71247	56	52.99576	88	83.27906
25	23.65882	57	53.94212	89	84.22541
26	24.60518	58	54.88847	90	85.17176
27	25.55153	59	55.83482	91	86.11812
28	26.49788	60	56.78118	92	87.06447
29	27.44424	61	57.72753	93	88.01082
30	28.39059	62	58.67388	94	88.95718
31	29.33694	63	59.62024	95	89.90353
32	30.28329	64	60.56659	96	90.84988

Table 19 *(continued)*

qt.	l.	qt.	l.	qt.	l.
97	91.79624	131	123.97223	165	156.14824
98	92.74259	132	124.91859	166	157.09459
99	93.68894	133	125.86494	167	158.04094
100	94.63529	134	126.81129	168	158.98729
101	95.58165	135	127.75765	169	159.93365
102	96.52800	136	128.70400	170	160.88000
103	97.47435	137	129.65035	171	161.82635
104	98.42071	138	130.59670	172	162.77271
105	99.36706	139	131.54306	173	163.71906
106	100.31341	140	132.48941	174	164.66541
107	101.25976	141	133.43576	175	165.61176
108	102.20612	142	134.38212	176	166.55812
109	103.15247	143	135.32847	177	167.50447
110	104.09882	144	136.27482	178	168.45082
111	105.04518	145	137.22118	179	169.39718
112	105.99153	146	138.16753	180	170.34353
113	106.93788	147	139.11388	181	171.28988
114	107.88424	148	140.06023	182	172.23623
115	108.83059	149	141.00659	183	173.18259
116	109.77694	150	141.95294	184	174.12894
117	110.72329	151	142.89929	185	175.07529
118	111.66965	152	143.84565	186	176.02165
119	112.61600	153	144.79200	187	176.96800
120	113.56235	154	145.73835	188	177.91435
121	114.50871	155	146.68471	189	178.86071
122	115.45506	156	147.63106	190	179.80706
123	116.40141	157	148.57741	191	180.75341
124	117.34776	158	149.52376	192	181.69976
125	118.29412	159	150.47012	193	182.64612
126	119.24047	160	151.41647	194	183.59247
127	120.18682	161	152.36282	195	184.53882
128	121.13318	162	153.30918	196	185.48518
129	122.07953	163	154.25553	197	186.43153
130	123.02588	164	155.20188	198	187.37788

Table 19 *(continued)*

qt.	l.	qt.	l.	qt.	l.
199	188.32423	300	283.90588	410	388.00470
200	189.27059	310	293.36941	420	397.46824
210	198.73412	320	302.83294	430	406.93176
220	208.19765	330	312.29647	440	416.39529
230	217.66118	340	321.76000	450	425.85882
240	227.12471	350	331.22353	460	435.32235
250	236.58824	360	340.68706	470	444.78588
260	246.05176	370	350.15059	480	454.24941
270	255.51529	380	359.61412	490	463.71294
280	264.97882	390	369.07764	500	473.17647
290	274.44235	400	378.54118		

Table 20 Gallons to Liters

1 gallon = 3.7854 liters

gal.	l.	gal.	l.	gal.	l.
1	3.7854	33	124.9186	65	246.0518
2	7.5708	34	128.7040	66	249.8372
3	11.3562	35	132.4894	67	253.6226
4	15.1416	36	136.2748	68	257.4080
5	18.9271	37	140.0602	69	261.1934
6	22.7125	38	143.8456	70	264.9788
7	26.4979	39	147.6311	71	268.7642
8	30.2833	40	151.4165	72	272.5496
9	34.0687	41	155.2019	73	276.3351
10	37.8541	42	158.9873	74	280.1205
11	41.6395	43	162.7727	75	283.9059
12	45.4249	44	166.5581	76	287.6913
13	49.2104	45	170.3435	77	291.4767
14	52.9958	46	174.1289	78	295.2621
15	56.7812	47	177.9144	79	299.0475
16	60.5666	48	181.6998	80	302.8329
17	64.3520	49	185.4852	81	306.6184
18	68.1374	50	189.2706	82	310.4038
19	71.9228	51	193.0560	83	314.1892
20	75.7082	52	196.8414	84	317.9746
21	79.4936	53	200.6268	85	321.7600
22	83.2791	54	204.4122	86	325.5454
23	87.0645	55	208.1976	87	329.3308
24	90.8499	56	211.9831	88	333.1162
25	94.6353	57	215.7685	89	336.9016
26	98.4207	58	219.5539	90	340.6871
27	102.2061	59	223.3393	91	344.4725
28	105.9915	60	227.1247	92	348.2579
29	109.7769	61	230.9101	93	352.0433
30	113.5624	62	234.6955	94	355.8287
31	117.3478	63	238.4809	95	359.6141
32	121.1332	64	242.2664	96	363.3995

Table 20 *(continued)*

gal.	l.	gal.	l.	gal.	l.
97	367.1849	131	495.8889	165	624.5929
98	370.9704	132	499.6744	166	628.3783
99	374.7558	133	503.4598	167	632.1638
100	378.5412	134	507.2452	168	635.9492
101	382.3266	135	511.0306	169	639.7346
102	386.1120	136	514.8160	170	643.5200
103	389.8974	137	518.6014	171	647.3054
104	393.6828	138	522.3868	172	651.0908
105	397.4682	139	526.1722	173	654.8762
106	401.2536	140	529.9576	174	658.6616
107	405.0391	141	533.7431	175	662.4471
108	408.8245	142	537.5285	176	666.2325
109	412.6099	143	541.3139	177	670.0179
110	416.3953	144	545.0993	178	673.8033
111	420.1807	145	548.8847	179	677.5887
112	423.9661	146	552.6701	180	681.3741
113	427.7515	147	556.4555	181	685.1595
114	431.5369	148	560.2409	182	688.9449
115	435.3224	149	564.0264	183	692.7304
116	439.1078	150	567.8118	184	696.5158
117	442.8932	151	571.5972	185	700.3012
118	446.6786	152	575.3826	186	704.0866
119	450.4640	153	579.1680	187	707.8720
120	454.2494	154	582.9534	188	711.6574
121	458.0348	155	586.7388	189	715.4428
122	461.8202	156	590.5242	190	719.2282
123	465.6056	157	594.3096	191	723.0136
124	469.3911	158	598.0951	192	726.7991
125	473.1765	159	601.8805	193	730.5845
126	476.9619	160	605.6659	194	734.3699
127	480.7473	161	609.4513	195	738.1553
128	484.5327	162	613.2367	196	741.9407
129	488.3181	163	617.0221	197	745.7261
130	492.1035	164	620.8075	198	749.5115

Table 20 *(continued)*

gal.	l.	gal.	l.	gal.	l.
199	753.2969	233	882.0009	267	1010.7049
200	757.0824	234	885.7864	268	1014.4903
201	760.8678	235	889.5718	269	1018.2758
202	764.6532	236	893.3572	270	1022.0612
203	768.4386	237	897.1426	271	1025.8466
204	772.2240	238	900.9280	272	1029.6320
205	776.0094	239	904.7134	273	1033.4174
206	779.7948	240	908.4988	274	1037.2028
207	783.5802	241	912.2842	275	1040.9882
208	787.3656	242	916.0696	276	1044.7736
209	791.1511	243	919.8551	277	1048.5591
210	794.9365	244	923.6405	278	1052.3445
211	798.7219	245	927.4259	279	1056.1299
212	802.5073	246	931.2113	280	1059.9153
213	806.2927	247	934.9967	281	1063.7007
214	810.0781	248	938.7821	282	1067.4861
215	813.8635	249	942.5675	283	1071.2715
216	817.6489	250	946.3529	284	1075.0569
217	821.4343	251	950.1384	285	1078.8423
218	825,2198	252	953.9238	286	1082.6278
219	829.0052	253	957.7092	287	1086.4132
220	832.7906	254	961.4946	288	1090.1986
221	836.5760	255	965.2800	289	1093.9840
222	840.3614	256	969.0654	290	1097.7694
223	844.1468	257	972.8508	291	1101.5548
224	847.9322	258	976.6362	292	1105.3402
225	851.7176	259	980.4216	293	1109.1256
226	855.5031	260	984.2071	294	1112.9111
227	859.2885	261	987.9925	295	1116.6965
228	863.0739	262	991.7779	296	1120.4819
229	866.8593	263	995.5633	297	1124.2673
230	870.6447	264	999.3487	298	1128.0527
231	874.4301	265	1003.1341	299	1131.8381
232	878.2155	266	1006.9195	300	1135.6235

Table 20 *(continued)*

gal.	l.	gal.	l.	gal.	l.
301	1139.4089	335	1268.1129	369	1396.8169
302	1143.1944	336	1271.8983	370	1400.6024
303	1146.9798	337	1275.6838	371	1404.3878
304	1150.7652	338	1279.4692	372	1408.1732
305	1154.5506	339	1283.2546	373	1411.9586
306	1158.3360	340	1287.0400	374	1415.7440
307	1162.1214	341	1290.8254	375	1419.5294
308	1165.9068	342	1294.6108	376	1423.3148
309	1169.6922	343	1298.3962	377	1427.1002
310	1173.4776	344	1302.1816	378	1430.8857
311	1177.2631	345	1305.9671	379	1434.6711
312	1181.0485	346	1309.7525	380	1438.4565
313	1184.8339	347	1313.5379	381	1442.2419
314	1188.6193	348	1317.3233	382	1446.0273
315	1192.4047	349	1321.1087	383	1449.8127
316	1196.1901	350	1324.8941	384	1453.5981
317	1199.9755	351	1328.6795	385	1457.3835
318	1203.7609	352	1332.4649	386	1461.1689
319	1207.5464	353	1336.2504	387	1464.9543
320	1211.3318	354	1340.0358	388	1468.7398
321	1215.1172	355	1343.8212	389	1472.5252
322	1218.9026	356	1347.6066	390	1476.3106
323	1222.6880	357	1351.3920	391	1480.0960
324	1226.4734	358	1355.1774	392	1483.8814
325	1230.2588	359	1358.9628	393	1487.6668
326	1234.0442	360	1362.7482	394	1491.4522
327	1237.8296	361	1366.5336	395	1495.2376
328	1241.6151	362	1370.3191	396	1499.0231
329	1245.4005	363	1374.1045	397	1502.8085
330	1249.1859	364	1377.8899	398	1506.5939
331	1252.9713	365	1381.6753	399	1510.3793
332	1256.7567	366	1385.4607	400	1514.1647
333	1260.5421	367	1389.2461	401	1517.9501
334	1264.3275	368	1393.0315	402	1521.7355

Table 20 *(continued)*

gal.	l.	gal.	l.	gal.	l.
403	1525.5209	437	1654.2249	471	1782.9289
404	1529.3064	438	1658.0103	472	1786.7144
405	1533.0918	439	1661.7958	473	1790.4998
406	1536.8772	440	1665.5812	474	1794.2852
407	1540.6626	441	1669.3666	475	1798.0706
408	1544.4480	442	1673.1520	476	1801.8560
409	1548.2334	443	1676.9374	477	1805.6414
410	1552.0188	444	1680.7228	478	1809.4268
411	1555.8042	445	1684.5082	479	1813.2122
412	1559.5896	446	1688.2936	480	1816.9977
413	1563.3751	447	1692.0791	481	1820.7831
414	1567.1605	448	1695.8645	482	1824.5685
415	1570.9459	449	1699.6499	483	1828.3539
416	1574.7313	450	1703.4353	484	1832.1393
417	1578.5167	451	1707.2207	485	1835.9247
418	1582.3021	452	1711.0061	486	1839.7101
419	1586.0875	453	1714.7915	487	1843.4955
420	1589.8729	454	1718.5769	488	1847.2809
421	1593.6584	455	1722.3624	489	1851.0663
422	1597.4438	456	1726.1478	490	1854.8518
423	1601.2292	457	1729.9332	491	1858.6372
424	1605.0146	458	1733.7186	492	1862.4226
425	1608.8000	459	1737.5040	493	1866.2080
426	1612.5854	460	1741.2894	494	1869.9934
427	1616.3708	461	1745.0748	495	1873.7788
428	1620.1562	462	1748.8602	496	1877.5642
429	1623.9417	463	1752.6456	497	1881.3496
430	1627.7271	464	1756.4311	498	1885.1351
431	1631.5125	465	1760.2165	499	1888.9205
432	1635.2979	466	1764.0019	500	1892.7059
433	1639.0833	467	1767.7873	510	1930.5600
434	1642.8687	468	1771.5727	520	1968.4141
435	1646.6541	469	1775.3581	530	2006.2682
436	1650.4395	470	1779.1435	540	2044.1223

Table 20 *(continued)*

gal.	l.	gal.	l.	gal.	l.
550	2081.9765	710	2687.6423	860	3255.4541
560	2119.8306	720	2725.4965	870	3293.3082
570	2157.6847	730	2763.3506	880	3331.1624
580	2195.5388	740	2801.2047	890	3369.0165
590	2233.3929	750	2839.0588	900	3406.8706
600	2271.2470	760	2876.9129	910	3444.7247
610	2309.1012	770	2914.7671	920	3482.5788
620	2346.9553	780	2952.6212	930	3520.4329
630	2384.8094	790	2990.4753	940	3558.2870
640	2422.6635	800	3028.3294	950	3596.1412
650	2460.5176	810	3066.1835	960	3633.9953
660	2498.3718	820	3104.0376	970	3671.8494
670	2536.2259	830	3141.8918	980	3709.7035
680	2574.0800	840	3179.7459	990	3747.5576
690	2611.9341	850	3217.6000	1000	3785.4120
700	2649.7882				

Table 21 Bushels to Liters

1 bushel = 35.2391 liters

bu.	l.	bu.	l.	bu.	l.
¼	8.810	30	1057.172	62	2184.822
½	17.620	31	1092.411	63	2220.061
¾	26.429	32	1127.650	64	2255.300
1	35.239	33	1162.889	65	2290.540
2	70.478	34	1198.128	66	2325.779
3	105.717	35	1233.367	67	2361.018
4	140.956	36	1268.607	68	2396.257
5	176.195	37	1303.846	69	2431.496
6	211.434	38	1339.085	70	2466.735
7	246.673	39	1374.324	71	2501.974
8	281.913	40	1409.563	72	2537.213
9	317.152	41	1444.802	73	2572.452
10	352.391	42	1480.041	74	2607.691
11	387.630	43	1515.280	75	2642.930
12	422.869	44	1550.519	76	2678.169
13	458.108	45	1585.758	77	2713.408
14	493.347	46	1620.997	78	2748.647
15	528.586	47	1656.236	79	2783.887
16	563.825	48	1691.475	80	2819.126
17	599.064	49	1726.714	81	2854.365
18	634.303	50	1761.953	82	2889.604
19	669.542	51	1797.193	83	2924.843
20	704.781	52	1832.432	84	2960.082
21	740.020	53	1867.671	85	2995.321
22	775.260	54	1902.910	86	3030.560
23	810.499	55	1938.149	87	3065.799
24	845.738	56	1973.388	88	3101.038
25	880.977	57	2008.627	89	3136.277
26	916.216	58	2043.866	90	3171.516
27	951.455	59	2079.105	91	3206.755
28	986.694	60	2114.344	92	3241.994
29	1021.933	61	2149.583	93	3277.233

Table 21 *(continued)*

bu.	l.	bu.	l.	bu.	l.
94	3312.473	128	4510.601	162	5708.729
95	3347.712	129	4545.840	163	5743.968
96	3382.951	130	4581.079	164	5779.207
97	3418.190	131	4616.318	165	5814.447
98	3453.429	132	4651.557	166	5849.686
99	3488.668	133	4686.796	167	5884.925
100	3523.907	134	4722.035	168	5920.164
101	3559.146	135	4757.274	169	5955.403
102	3594.385	136	4792.513	170	5990.642
103	3629.624	137	4827.753	171	6025.881
104	3664.863	138	4862.992	172	6061.120
105	3700.102	139	4898.231	173	6096.359
106	3735.341	140	4933.470	174	6131.598
107	3770.580	141	4968.709	175	6166.837
108	3805.820	142	5003.948	176	6202.076
109	3841.059	143	5039.187	177	6237.315
110	3876.298	144	5074.426	178	6272.554
111	3911.537	145	5109.665	179	6307.794
112	3946.776	146	5144.904	180	6343.033
113	3982.015	147	5180.143	181	6378.272
114	4017.254	148	5215.382	182	6413.511
115	4052.493	149	5250.621	183	6448.750
116	4087.732	150	5285.860	184	6483.989
117	4122.971	151	5321.100	185	6519.228
118	4158.210	152	5356.339	186	6554.467
119	4193.449	153	5391.578	187	6589.706
120	4228.688	154	5426.817	188	6624.945
121	4263.927	155	5462.056	189	6660.184
122	4299.167	156	5497.295	190	6695.423
123	4334.406	157	5532.534	191	6730.662
124	4369.645	158	5567.773	192	6765.901
125	4404.884	159	5603.012	193	6801.140
126	4440.123	160	5638.251	194	6836.380
127	4475.362	161	5673.490	195	6871.619

Table 21 *(continued)*

bu.	l.	bu.	l.	bu.	l.
196	6906.858	300	10571.721	430	15152.800
197	6942.097	310	10924.112	440	15505.191
198	6977.336	320	11276.502	450	15857.581
199	7012.575	330	11628.893	460	16209.972
200	7047.814	340	11981.284	470	16562.363
210	7400.205	350	12333.674	480	16914.753
220	7752.595	360	12686.065	490	17267.144
230	8104.986	370	13038.456	500	17619.550
240	8457.377	380	13390.847	600	21143.44
250	8809.767	390	13743.237	700	24667.35
260	9162.158	400	14095.628	800	28191.26
270	9514.549	410	14448.019	900	31715.16
280	9866.940	420	14800.409	1000	35239.07
290	10219.330				

Volume

Table 22 Cubic Inches to Cubic Centimeters

1 cubic inch = 16.3871 cubic centimeters

in.^{3}	cm.^{3}	in.^{3}	cm.^{3}	in.^{3}	cm.^{3}
1/8	2.048	26	426.064	58	950.450
1/4	4.097	27	442.451	59	966.837
3/8	6.145	28	458.838	60	983.224
1/2	8.194	29	475.225	61	999.611
5/8	10.242	30	491.612	62	1015.998
3/4	12.290	31	507.999	63	1032.385
7/8	14.339	32	524.386	64	1048.772
1	16.387	33	540.773	65	1065.159
2	32.774	34	557.160	66	1081.546
3	49.161	35	573.547	67	1097.933
4	65.548	36	589.934	68	1114.320
5	81.935	37	606.321	69	1130.707
6	98.322	38	622.708	70	1147.094
7	114.709	39	639.095	71	1163.482
8	131.097	40	655.483	72	1179.869
9	147.484	41	671.870	73	1196.256
10	163.871	42	688.257	74	1212.643
11	180.258	43	704.644	75	1229.030
12	196.645	44	721.031	76	1245.417
13	213.032	45	737.418	77	1261.804
14	229.419	46	753.805	78	1278.191
15	245.806	47	770.192	79	1294.578
16	262.193	48	786.579	80	1310.965
17	278.580	49	802.966	81	1327.352
18	294.967	50	819.353	82	1343.739
19	311.354	51	835.740	83	1360.126
20	327.741	52	852.127	84	1376.513
21	344.128	53	868.514	85	1392.900
22	360.515	54	884.901	86	1409.287
23	376.902	55	901.289	87	1425.675
24	393.290	56	917.676	88	1442.062
25	409.677	57	934.063	89	1458.449

Table 22 *(continued)*

in.^{3}	cm.^{3}	in.^{3}	cm.^{3}	in.^{3}	cm.^{3}
90	1474.836	124	2031.996	158	2589.156
91	1491.223	125	2048.383	159	2605.543
92	1507.610	126	2064.770	160	2621.930
93	1523.997	127	2081.157	161	2638.317
94	1540.384	128	2097.544	162	2654.704
95	1556.771	129	2113.931	163	2671.091
96	1573.158	130	2130.318	164	2687.478
97	1589.545	131	2146.705	165	2703.866
98	1605.932	132	2163.092	166	2720.253
99	1622.319	133	2179.479	167	2736.640
100	1638.706	134	2195.867	168	2753.027
101	1655.093	135	2212.254	169	2769.414
102	1671.481	136	2228.641	170	2785.801
103	1687.868	137	2245.028	171	2802.188
104	1704.255	138	2261.415	172	2818.575
105	1720.642	139	2277.802	173	2834.962
106	1737.029	140	2294.189	174	2851.349
107	1753.416	141	2310.576	175	2867.736
108	1769.803	142	2326.963	176	2884.123
109	1786.190	143	2343.350	177	2900.510
110	1802.577	144	2359.737	178	2916.897
111	1818.964	145	2376.124	179	2933.284
112	1835.351	146	2392.511	180	2949.672
113	1851.738	147	2408.898	181	2966.059
114	1868.125	148	2425.285	182	2982.446
115	1884.512	149	2441.673	183	2998.833
116	1900.899	150	2458.060	184	3015.220
117	1917.286	151	2474.447	185	3031.607
118	1933.674	152	2490.834	186	3047.994
119	1950.061	153	2507.221	187	3064.381
120	1966.448	154	2523.608	188	3080.768
121	1982.835	155	2539.995	189	3097.155
122	1999.222	156	2556.382	190	3113.542
123	2015.609	157	2572.769	191	3129.929

Table 22 *(continued)*

in.^{3}	cm.^{3}	in.^{3}	cm.^{3}	in.^{3}	cm.^{3}
192	3146.316	250	4096.766	380	6227.084
193	3162.703	260	4260.637	390	6390.955
194	3179.090	270	4424.507	400	6554.826
195	3195.477	280	4588.378	410	6718.696
196	3211.865	290	4752.249	420	6882.567
197	3228.252	300	4916.119	430	7046.438
198	3244.639	310	5079.990	440	7210.308
199	3261.026	320	5243.860	450	7374.179
200	3277.413	330	5407.731	460	7538.049
210	3441.283	340	5571.602	470	7701.920
220	3605.154	350	5735.472	480	7865.791
230	3769.025	360	5899.343	490	8029.661
240	3932.895	370	6063.214	500	8193.532

Table 23 Cubic Feet to Cubic Meters

1 cubic foot = .0283 cubic meter

$ft.^3$	$m.^3$	$ft.^3$	$m.^3$	$ft.^3$	$m.^3$
1	0.028317	33	0.934456	65	1.840595
2	0.056634	34	0.962773	66	1.868912
3	0.084951	35	0.991090	67	1.897229
4	0.113267	36	1.019406	68	1.925546
5	0.141584	37	1.047723	69	1.953862
6	0.169901	38	1.076040	70	1.982179
7	0.198218	39	1.104357	71	2.010496
8	0.226535	40	1.132674	72	2.038813
9	0.254852	41	1.160991	73	2.067130
10	0.283168	42	1.189308	74	2.095447
11	0.311485	43	1.217624	75	2.123763
12	0.339802	44	1.245941	76	2.152080
13	0.368119	45	1.274258	77	2.180397
14	0.396436	46	1.302575	78	2.208714
15	0.424753	47	1.330892	79	2.237031
16	0.453070	48	1.359209	80	2.265348
17	0.481386	49	1.387525	81	2.293665
18	0.509703	50	1.415842	82	2.321981
19	0.538020	51	1.444159	83	2.350298
20	0.566337	52	1.472476	84	2.378615
21	0.594654	53	1.500793	85	2.406932
22	0.622971	54	1.529110	86	2.435249
23	0.651287	55	1.557427	87	2.463566
24	0.679604	56	1.585743	88	2.491882
25	0.707921	57	1.614060	89	2.520199
26	0.736238	58	1.642377	90	2.548516
27	0.764555	59	1.670694	91	2.576833
28	0.792872	60	1.699011	92	2.605150
29	0.821189	61	1.727328	93	2.633467
30	0.849505	62	1.755644	94	2.661784
31	0.877822	63	1.783961	95	2.690100
32	0.906139	64	1.812278	96	2.718417

Table 23 *(continued)*

ft.^{3}	m.^{3}	ft.^{3}	m.^{3}	ft.^{3}	m.^{3}
97	2.746734	131	3.709507	165	4.672280
98	2.775051	132	3.737824	166	4.700597
99	2.803368	133	3.766141	167	4.728913
100	2.831685	134	3.794457	168	4.757230
101	2.860002	135	3.822774	169	4.785547
102	2.888318	136	3.851091	170	4.813864
103	2.916635	137	3.879408	171	4.842181
104	2.944952	138	3.907725	172	4.870498
105	2.973269	139	3.936042	173	4.898814
106	3.001586	140	3.964359	174	4.927131
107	3.029903	141	3.992675	175	4.955448
108	3.058219	142	4.020992	176	4.983765
109	3.086536	143	4.049309	177	5.012082
110	3.114853	144	4.077626	178	5.040399
111	3.143170	145	4.105943	179	5.068716
112	3.171487	146	4.134260	180	5.097032
113	3.199804	147	4.162576	181	5.125349
114	3.228121	148	4.190893	182	5.153666
115	3.256437	149	4.219210	183	5.181983
116	3.284754	150	4.247527	184	5.210300
117	3.313071	151	4.275844	185	5.238617
118	3.341388	152	4.304161	186	5.266933
119	3.369705	153	4.332478	187	5.295250
120	3.398022	154	4.360794	188	5.323567
121	3.426338	155	4.389111	189	5.351884
122	3.454655	156	4.417428	190	5.380201
123	3.482972	157	4.445745	191	5.408518
124	3.511289	158	4.474062	192	5.436835
125	3.539606	159	4.502379	193	5.465151
126	3.567923	160	4.530695	194	5.493468
127	3.596240	161	4.559012	195	5.521785
128	3.624556	162	4.587329	196	5.550102
129	3.652873	163	4.615646	197	5.578419
130	3.681190	164	4.643963	198	5.606736

Table 23 *(continued)*

ft.^{3}	m.^{3}	ft.^{3}	m.^{3}	ft.^{3}	m.^{3}
199	5.635052	520	14.724760	850	24.069319
200	5.663369	530	15.007929	860	24.352488
210	5.946538	540	15.291097	870	24.635656
220	6.229706	550	15.574266	880	24.918825
230	6.512875	560	15.857434	890	25.201993
240	6.796043	570	16.140602	900	25.485162
250	7.079212	580	16.423771	910	25.768330
260	7.362380	590	16.706939	920	26.051499
270	7.645549	600	16.990108	930	26.334667
280	7.928717	610	17.273276	940	26.617836
290	8.211885	620	17.556445	950	26.901004
300	8.495054	630	17.839613	960	27.184173
310	8.778222	640	18.122782	970	27.467341
320	9.061391	650	18.405950	980	27.750510
330	9.344559	660	18.689119	990	28.033678
340	9.627728	670	18.972287	1000	28.31685
350	9.910896	680	19.255455	2000	56.63369
360	10.194065	690	19.538624	3000	84.95054
370	10.477233	700	19.821793	4000	113.267
380	10.760402	710	20.104961	5000	141.584
390	11.043570	720	20.388129	6000	169.901
400	11.326739	730	20.671298	7000	198.218
410	11.609907	740	20.954466	8000	226.535
420	11.893075	750	21.237635	9000	254.852
430	12.176244	760	21.520803	10000	283.168
440	12.459412	770	21.803972	15000	424.753
450	12.742581	780	22.087140	20000	566.337
460	13.025749	790	22.370309	25000	707.921
470	13.308918	800	22.653477	30000	849.505
480	13.592086	810	22.936646	35000	991.090
490	13.875255	820	23.219814	40000	1132.674
500	14.158423	830	23.502983	45000	1274.258
510	14.441592	840	23.786151	50000	1415.842

Table 24 Cubic Yards to Cubic Meters

1 cubic yard = .7645 cubic meter

yd.^{3}	m.^{3}	yd.^{3}	m.^{3}	yd.^{3}	m.^{3}
1	0.76455	33	25.23031	65	49.69607
2	1.52911	34	25.99487	66	50.46062
3	2.29366	35	26.75942	67	51.22518
4	3.05822	36	27.52397	68	51.98973
5	3.82277	37	28.28853	69	52.75428
6	4.58733	38	29.05308	70	53.51884
7	5.35188	39	29.81764	71	54.28339
8	6.11644	40	30.58219	72	55.04795
9	6.88099	41	31.34675	73	55.81250
10	7.64555	42	32.11130	74	56.57706
11	8.41010	43	32.87586	75	57.34161
12	9.17466	44	33.64041	76	58.10617
13	9.93921	45	34.40497	77	58.87072
14	10.70377	46	35.16952	78	59.63528
15	11.46832	47	35.93408	79	60.39983
16	12.23288	48	36.69863	80	61.16439
17	12.99743	49	37.46319	81	61.92894
18	13.76199	50	38.22774	82	62.69350
19	14.52654	51	38.99230	83	63.45805
20	15.29110	52	39.75685	84	64.22261
21	16.05565	53	40.52141	85	64.98716
22	16.82021	54	41.28596	86	65.75172
23	17.58476	55	42.05052	87	66.51627
24	18.34932	56	42.81507	88	67.28083
25	19.11387	57	43.57963	89	68.04538
26	19.87843	58	44.34418	90	68.80994
27	20.64298	59	45.10874	91	69.57449
28	21.40754	60	45.87329	92	70.33905
29	22.17209	61	46.63785	93	71.10360
30	22.93665	62	47.40240	94	71.86816
31	23.70120	63	48.16696	95	72.63271
32	24.46576	64	48.93151	96	73.39727

Table 24 *(continued)*

yd.^{3}	m.^{3}	yd.^{3}	m.^{3}	yd.^{3}	m.^{3}
97	74.16182	131	100.15669	165	126.15155
98	74.92638	132	100.92124	166	126.91611
99	75.69093	133	101.68580	167	127.68066
100	76.45549	134	102.45035	168	128.44522
101	77.22004	135	103.21491	169	129.20977
102	77.98460	136	103.97946	170	129.97433
103	78.74915	137	104.74401	171	130.73888
104	79.51371	138	105.50857	172	131.50344
105	80.27826	139	106.27312	173	132.26799
106	81.04281	140	107.03768	174	133.03255
107	81.80737	141	107.80223	175	133.79710
108	82.57192	142	108.56679	176	134.56166
109	83.33648	143	109.33134	177	135.32621
110	84.10103	144	110.09590	178	136.09076
111	84.86559	145	110.86045	179	136.85532
112	85.63014	146	111.62501	180	137.61987
113	86.39470	147	112.38956	181	138.38443
114	87.15925	148	113.15412	182	139.14898
115	87.92381	149	113.91867	183	139.91354
116	88.68836	150	114.68323	184	140.67809
117	89.45292	151	115.44778	185	141.44265
118	90.21747	152	116.21234	186	142.20720
119	90.98203	153	116.97689	187	142.97176
120	91.74658	154	117.74145	188	143.73631
121	92.51114	155	118.50600	189	144.50087
122	93.27569	156	119.27056	190	145.26542
123	94.04025	157	120.03511	191	146.02998
124	94.80480	158	120.79967	192	146.79453
125	95.56936	159	121.56422	193	147.55909
126	96.33391	160	122.32878	194	148.32364
127	97.09847	161	123.09333	195	149.08820
128	97.86302	162	123.85789	196	149.85275
129	98.62758	163	124.62244	197	150.61731
130	99.39213	164	125.38700	198	151.38186

Table 24 *(continued)*

yd.^{3}	m.^{3}	yd.^{3}	m.^{3}	yd.^{3}	m.^{3}
199	152.14642	233	178.14128	267	204.13615
200	152.91097	234	178.90584	268	204.90070
201	153.67553	235	179.67039	269	205.66526
202	154.44008	236	180.43495	270	206.42981
203	155.20464	237	181.19950	271	207.19437
204	155.96919	238	181.96406	272	207.95892
205	156.73375	239	182.72861	273	208.72347
206	157.49830	240	183.49317	274	209.48803
207	158.26286	241	184.25772	275	210.25258
208	159.02741	242	185.02227	276	211.01714
209	159.79196	243	185.78683	277	211.78169
210	160.55652	244	186.55138	278	212.54625
211	161.32107	245	187.31594	279	213.31080
212	162.08563	246	188.08049	280	214.07536
213	162.85018	247	188.84505	281	214.83991
214	163.61474	248	189.60960	282	215.60447
215	164.37929	249	190.37416	283	216.36902
216	165.14385	250	191.13871	284	217.13358
217	165.90840	251	191.90327	285	217.89813
218	166.67296	252	192.66782	286	218.66269
219	167.43751	253	193.43238	287	219.42724
220	168.20207	254	194.19693	288	220.19180
221	168.96662	255	194.96149	289	220.95635
222	169.73118	256	195.72604	290	221.72091
223	170.49573	257	196.49060	291	222.48546
224	171.26029	258	197.25515	292	223.25002
225	172.02484	259	198.01971	293	224.01457
226	172.78940	260	198.78426	294	224.77913
227	173.55395	261	199.54882	295	225.54368
228	174.31851	262	200.31337	296	226.30824
229	175.08306	263	201.07793	297	227.07279
230	175.84762	264	201.84248	298	227.83735
231	176.61217	265	202.60704	299	228.60190
232	177.37673	266	203.37159	300	229.36646

Table 24 *(continued)*

yd.^{3}	m.^{3}	yd.^{3}	m.^{3}	yd.^{3}	m.^{3}
301	230.13101	335	256.12588	369	282.12074
302	230.89557	336	256.89043	370	282.88530
303	231.66012	337	257.65498	371	283.64985
304	232.42468	338	258.41954	372	284.41441
305	233.18923	339	259.18409	373	285.17896
306	233.95378	340	259.94865	374	285.94352
307	234.71834	341	260.71320	375	286.70807
308	235.48289	342	261.47776	376	287.47263
309	236.24745	343	262.24231	377	288.23718
310	237.01200	344	263.00687	378	289.00174
311	237.77656	345	263.77142	379	289.76629
312	238.54111	346	264.53598	380	290.53085
313	239.30567	347	265.30053	381	291.29540
314	240.07022	348	266.06509	382	292.05996
315	240.83478	349	266.82964	383	292.82451
316	241.59933	350	267.59420	384	293.58907
317	242.36389	351	268.35875	385	294.35362
318	243.12844	352	269.12331	386	295.11817
319	243.89300	353	269.88786	387	295.88273
320	244.65755	354	270.65242	388	296.64728
321	245.42211	355	271.41697	389	297.41184
322	246.18666	356	272.18153	390	298.17639
323	246.95122	357	272.94608	391	298.94095
324	247.71577	358	273.71064	392	299.70550
325	248.48033	359	274.47519	393	300.47006
326	249.24488	360	275.23975	394	301.23461
327	250.00944	361	276.00430	395	301.99917
328	250.77399	362	276.76886	396	302.76372
329	251.53855	363	277.53341	397	303.52828
330	252.30310	364	278.29797	398	304.29283
331	253.06766	365	279.06252	399	305.05739
332	253.83221	366	279.82708	400	305.82194
333	254.59677	367	280.59163	401	306.58649
334	255.36132	368	281.35619	402	307.35105

Table 24 *(continued)*

yd.^{3}	m.^{3}	yd.^{3}	m.^{3}	yd.^{3}	m.^{3}
403	308.11560	437	334.11047	471	360.10534
404	308.88016	438	334.87503	472	360.86989
405	309.64471	439	335.63958	473	361.63445
406	310.40927	440	336.40414	474	362.39900
407	311.17382	441	337.16869	475	363.16356
408	311.93838	442	337.93325	476	363.92811
409	312.70293	443	338.69780	477	364.69267
410	313.46749	444	339.46236	478	365.45722
411	314.23204	445	340.22691	479	366.22178
412	314.99660	446	340.99147	480	366.98633
413	315.76115	447	341.75602	481	367.75089
414	316.52571	448	342.52058	482	368.51544
415	317.29026	449	343.28513	483	369.27999
416	318.05482	450	344.04968	484	370.04455
417	318.81937	451	344.81424	485	370.80910
418	319.58393	452	345.57879	486	371.57366
419	320.34848	453	346.34335	487	372.33821
420	321.11304	454	347.10790	488	373.10277
421	321.87759	455	347.87246	489	373.86732
422	322.64215	456	348.63701	490	374.63188
423	323.40670	457	349.40157	491	375.39643
424	324.17126	458	350.16612	492	376.16099
425	324.93581	459	350.93068	493	376.92554
426	325.70037	460	351.69523	494	377.69010
427	326.46492	461	352.45979	495	378.45465
428	327.22948	462	353.22434	496	379.21921
429	327.99403	463	353.98890	497	379.98376
430	328.75859	464	354.75345	498	380.74832
431	329.52314	465	355.51801	499	381.51287
432	330.28770	466	356.28256	500	382.27743
433	331.05225	467	357.04712	510	389.92298
434	331.81681	468	357.81167	520	397.56852
435	332.58136	469	358.57623	530	405.21407
436	333.34592	470	359.34078	540	412.85962

Table 24 *(continued)*

yd.^{3}	m.^{3}	yd.^{3}	m.^{3}	yd.^{3}	m.^{3}
550	420.50517	760	581.06169	970	741.61821
560	428.15072	770	588.70724	980	749.26376
570	435.79627	780	596.35278	990	756.90931
580	443.44181	790	603.99834	1000	764.555
590	451.08736	800	611.64388	2000	1529.110
600	458.73291	810	619.28943	3000	2293.665
610	466.37846	820	626.93498	4000	3058.219
620	474.02401	830	634.58053	5000	3822.774
630	481.66956	840	642.22607	6000	4587.329
640	489.31511	850	649.87163	7000	5351.884
650	496.96066	860	657.51717	8000	6116.439
660	504.60620	870	665.16272	9000	6880.994
670	512.25175	880	672.80827	10000	7645.549
680	519.89730	890	680.45382	15000	11468.323
690	527.54285	900	688.09937	20000	15291.097
700	535.18840	910	695.74492	25000	19113.871
710	542.83395	920	703.39046	30000	22936.646
720	550.47949	930	711.03601	35000	26759.420
730	558.12505	940	718.68156	40000	30582.194
740	565.77059	950	726.32711	45000	34404.969
750	573.41614	960	733.97266	50000	38227.743

Temperature

Table 25 Degrees Fahrenheit to Degrees Celsius

1 degree Fahrenheit = .56 degree Celsius

F	C	F	C	F	C
0	−17.78	32	0.	64	17.78
1	−17.22	33	.56	65	18.33
2	−16.67	34	1.11	66	18.89
3	−16.11	35	1.67	67	19.44
4	−15.56	36	2.22	68	20.00
5	−15.00	37	2.78	69	20.56
6	−14.44	38	3.33	70	21.11
7	−13.89	39	3.89	71	21.67
8	−13.33	40	4.44	72	22.22
9	−12.78	41	5.00	73	22.78
10	−12.22	42	5.56	74	23.33
11	−11.67	43	6.11	75	23.89
12	−11.11	44	6.67	76	24.44
13	−10.56	45	7.22	77	25.00
14	−10.00	46	7.78	78	25.56
15	− 9.44	47	8.33	79	26.11
16	− 8.89	48	8.89	80	26.67
17	− 8.33	49	9.44	81	27.22
18	− 7.78	50	10.00	82	27.78
19	− 7.22	51	10.56	83	28.33
20	− 6.67	52	11.11	84	28.89
21	− 6.11	53	11.67	85	29.44
22	− 5.56	54	12.22	86	30.00
23	− 5.00	55	12.78	87	30.56
24	− 4.44	56	13.33	88	31.11
25	− 3.89	57	13.89	89	31.67
26	− 3.33	58	14.44	90	32.22
27	− 2.78	59	15.00	91	32.78
28	− 2.22	60	15.56	92	33.33
29	− 1.67	61	16.11	93	33.89
30	− 1.11	62	16.67	94	34.44
31	− .56	63	17.22	95	35.00

Table 25 *(continued)*

F	C	F	C	F	C
96	35.56	118	47.78	141	60.56
97	36.11	119	48.33	142	61.11
98	36.67	120	48.89	143	61.67
98.6	37.00	121	49.44	144	62.22
99	37.22	122	50.00	145	62.78
100	37.78	123	50.56	146	63.33
101	38.33	124	51.11	147	63.89
102	38.89	125	51.67	148	64.44
103	39.44	126	52.22	149	65.00
104	40.00	127	52.78	150	65.56
105	40.56	128	53.33	155	68.33
106	41.11	129	53.89	160	71.11
107	41.67	130	54.44	165	73.89
108	42.22	131	55.00	170	76.67
109	42.78	132	55.56	175	79.44
110	43.33	133	56.11	180	82.22
111	43.89	134	56.67	185	85.00
112	44.44	135	57.22	190	87.78
113	45.00	136	57.78	195	90.56
114	45.56	137	58.33	200	93.33
115	46.11	138	58.89	205	96.11
116	46.67	139	59.44	210	98.99
117	47.22	140	60.00	212	100.00

	Metric (Degrees Celsius)	Customary (Degrees Fahrenheit)
Water freezing point	0	32
Water boiling point	100	212
Normal body temperature	37	98.6

Metric to Customary

Unit	Multiply by	And Add	To Find
Degrees Centigrade	9/5	32	Degrees Fahrenheit

Customary to Metric

Unit	Multiply by	And Subtract	To Find
Degrees Fahrenheit	5/9	32	Degrees Centigrade

Metric to Customary
Length

Table 26 Millimeters to Inches

1 millimeter = .039 inch

mm.	in.	mm.	in.	mm.	in.
0.01	0.00039	0.33	0.01299	0.65	0.02559
0.02	0.00079	0.34	0.01339	0.66	0.02598
0.03	0.00118	0.35	0.01378	0.67	0.02638
0.04	0.00157	0.36	0.01417	0.68	0.02677
0.05	0.00197	0.37	0.01457	0.69	0.02717
0.06	0.00236	0.38	0.01496	0.70	0.02756
0.07	0.00276	0.39	0.01535	0.71	0.02795
0.08	0.00315	0.40	0.01575	0.72	0.02835
0.09	0.00354	0.41	0.01614	0.73	0.02874
0.10	0.00394	0.42	0.01654	0.74	0.02913
0.11	0.00433	0.43	0.01693	0.75	0.02953
0.12	0.00472	0.44	0.01732	0.76	0.02992
0.13	0.00512	0.45	0.01772	0.77	0.03031
0.14	0.00551	0.46	0.01811	0.78	0.03071
0.15	0.00591	0.47	0.01850	0.79	0.03110
0.16	0.00630	0.48	0.01890	0.80	0.03150
0.17	0.00669	0.49	0.01929	0.81	0.03189
0.18	0.00709	0.50	0.01969	0.82	0.03228
0.19	0.00748	0.51	0.02008	0.83	0.03268
0.20	0.00787	0.52	0.02047	0.84	0.03307
0.21	0.00827	0.53	0.02087	0.85	0.03346
0.22	0.00866	0.54	0.02126	0.86	0.03386
0.23	0.00906	0.55	0.02165	0.87	0.03425
0.24	0.00945	0.56	0.02205	0.88	0.03465
0.25	0.00984	0.57	0.02244	0.89	0.03504
0.26	0.01024	0.58	0.02283	0.90	0.03543
0.27	0.01063	0.59	0.02323	0.91	0.03583
0.28	0.01102	0.60	0.02362	0.92	0.03622
0.29	0.01142	0.61	0.02402	0.93	0.03661
0.30	0.01181	0.62	0.02441	0.94	0.03701
0.31	0.01220	0.63	0.02480	0.95	0.03740
0.32	0.01260	0.64	0.02520	0.96	0.03780

Table 26 *(continued)*

mm.	in.	mm.	in.	mm.	in.
0.97	0.03819	32	1.25984	66	2.59843
0.98	0.03858	33	1.29921	67	2.63780
0.99	0.03898	34	1.33858	68	2.67717
1	0.03937	35	1.37795	69	2.71654
2	0.07874	36	1.41732	70	2.75591
3	0.11811	37	1.45669	71	2.79528
4	0.15748	38	1.49606	72	2.83465
5	0.19685	39	1.53543	73	2.87402
6	0.23622	40	1.57480	74	2.91339
7	0.27559	41	1.61417	75	2.95276
8	0.31496	42	1.65354	76	2.99213
9	0.35433	43	1.69291	77	3.03150
10	0.39370	44	1.73228	78	3.07087
11	0.43307	45	1.77165	79	3.11024
12	0.47244	46	1.81102	80	3.14961
13	0.51181	47	1.85039	81	3.18898
14	0.55118	48	1.88976	82	3.22835
15	0.59055	49	1.92913	83	3.26772
16	0.62992	50	1.96850	84	3.30709
17	0.66929	51	2.00787	85	3.34646
18	0.70866	52	2.04724	86	3.38583
19	0.74803	53	2.08661	87	3.4520
20	0.78740	54	2.12598	88	3.46457
21	0.82677	55	2.16535	89	3.50394
22	0.86614	56	2.20472	90	3.54331
23	0.90551	57	2.24409	91	3.58268
24	0.94488	58	2.28346	92	3.62205
25	0.98425	59	2.32283	93	3.66142
26	1.02362	60	2.36220	94	3.70079
27	1.06299	61	2.40157	95	3.74016
28	1.10236	62	2.44094	96	3.77953
29	1.14173	63	2.48031	97	3.81890
30	1.18110	64	2.51969	98	3.85827
31	1.22047	65	2.55906	99	3.89764

Table 27 Centimeters to Feet, Inches

1 centimeter = 0 feet, .394 inch

cm.	ft.	in.	cm.	ft.	in.	cm.	ft.	in.
1		.394	33	1	.99	65	2	1.59
2		.787	34	1	1.39	66	2	1.98
3		1.181	35	1	1.78	67	2	2.38
4		1.575	36	1	2.17	68	2	2.77
5		1.969	37	1	2.57	69	2	3.17
6		2.362	38	1	2.96	70	2	3.56
7		2.756	39	1	3.35	71	2	3.95
8		3.150	40	1	3.75	72	2	4.35
9		3.543	41	1	4.14	73	2	4.74
10		3.937	42	1	4.54	74	2	5.13
11		4.33	43	1	4.93	75	2	5.53
12		4.72	44	1	5.32	76	2	5.92
13		5.12	45	1	5.72	77	2	6.31
14		5.51	46	1	6.11	78	2	6.71
15		5.91	47	1	6.50	79	2	7.10
16		6.30	48	1	6.90	80	2	7.50
17		6.69	49	1	7.29	81	2	7.89
18		7.09	50	1	7.69	82	2	8.28
19		7.48	51	1	8.08	83	2	8.68
20		7.87	52	1	8.47	84	2	9.07
21		8.27	53	1	8.87	85	2	9.46
22		8.66	54	1	9.26	86	2	9.86
23		9.06	55	1	9.65	87	2	10.25
24		9.45	56	1	10.05	88	2	10.65
25		9.84	57	1	10.44	89	2	11.04
26		10.24	58	1	10.83	90	2	11.43
27		10.63	59	1	11.23	91	2	11.83
28		11.02	60	1	11.62	92	3	.22
29		11.42	61	2	.02	93	3	.61
30		11.81	62	2	.41	94	3	1.01
31	1	.20	63	2	.80	95	3	1.40
32	1	.60	64	2	1.20	96	3	1.80

Table 27 *(continued)*

cm.	ft.	in.	cm.	ft.	in.	cm.	ft.	in.
97	3	2.19	99	3	2.98	100	3	3.37
98	3	2.58						

Table 28 Meters to Feet

1 meter = 3.2808 feet

m.	ft.	m.	ft.	m.	ft.
1	3.2808	33	108.2677	65	213.2546
2	6.5617	34	111.5486	66	216.5354
3	9.8425	35	114.8294	67	219.8163
4	13.1234	36	118.1102	68	223.0971
5	16.4042	37	121.3911	69	226.3780
6	19.6850	38	124.6719	70	229.6588
7	22.9659	39	127.9528	71	232.9396
8	26.2467	40	131.2336	72	236.2205
9	29.5276	41	134.5144	73	239.5013
10	32.8084	42	137.7753	74	242.7822
11	36.0892	43	141.0761	75	246.0630
12	39.3701	44	144.3570	76	249.2328
13	42.6509	45	147.6378	77	252.6247
14	45.9318	46	150.9186	78	255.9055
15	49.2126	47	154.1995	79	259.1863
16	52.4934	48	157.4803	80	262.4672
17	55.7743	49	160.7612	81	265.7480
18	59.0551	50	164.0420	82	269.0289
19	62.3360	51	167.3228	83	272.3097
20	65.6168	52	170.6037	84	275.5905
21	68.8976	53	173.8845	85	278.8714
22	76.1785	54	177.1654	86	282.1522
23	75.4593	55	180.4462	87	285.4331
24	78.7402	56	183.7270	88	288.7139
25	82.0210	57	187.0079	89	291.9947
26	85.3018	58	190.2887	90	295.2756
27	88.5827	59	193.5696	91	298.5564
28	91.8635	60	196.8504	92	301.8373
29	95.1444	61	200.1312	93	305.1181
30	98.4252	62	203.4121	94	308.3989
31	101.7060	63	206.6929	95	311.6798
32	104.9869	64	209.9738	96	314.9606

Table 28 *(continued)*

m.	ft.	m.	ft.	m.	ft.
97	318.2415	99	324.8031	100	328.0840
98	321.5223				

Table 29 Meters to Yards, Feet, Inches

1 meter = 1 yard, 0 feet, 3.37 inches

m.	yd.	ft.	in.	m.	yd.	ft.	in.
1	1		3.37	33	36		3.21
2	2		6.74	34	37		6.58
3	3		10.11	35	38		9.95
4	4	1	1.48	36	39	1	1.32
5	5	1	4.85	37	40	1	4.69
6	6	1	8.22	38	41	1	8.06
7	7	1	11.59	39	42	1	11.43
8	8	2	2.96	40	43	2	2.80
9	9	2	6.33	41	44	2	6.17
10	10	2	9.70	42	45	2	9.54
11	12		1.07	43	47		.91
12	13		4.44	44	48		4.28
13	14		7.81	45	49		7.65
14	15		11.18	46	50		11.02
15	16	1	2.55	47	51	1	2.39
16	17	1	5.92	48	52	1	5.76
17	18	1	9.29	49	53	1	9.13
18	19	2	.66	50	54	2	.50
19	20	2	4.03	51	55	2	3.87
20	21	2	7.40	52	56	2	7.24
21	22	2	10.77	53	57	2	10.61
22	24		2.14	54	59		1.98
23	25		5.51	55	60		5.35
24	26		8.88	56	61		8.72
25	27	1	.25	57	62	1	.09
26	28	1	3.62	58	63	1	3.46
27	29	1	6.99	59	64	1	6.83
28	30	1	10.36	60	65	1	10.20
29	31	2	1.73	61	66	2	1.57
30	32	2	5.10	62	67	2	4.94
31	33	2	8.47	63	68	2	8.31
32	34	2	11.84	64	69	2	11.69

Table 29 *(continued)*

m.	yd.	ft.	in.	m.	yd.	ft.	in.
65	71		3.06	83	90	2	3.72
66	72		6.43	84	91	2	7.09
67	73		9.80	85	92	2	10.46
68	74	1	1.17	86	94		1.83
69	75	1	4.54	87	95		5.20
70	76	1	7.91	88	96		8.57
71	77	1	11.28	89	97		11.94
72	78	2	2.65	90	98	1	3.31
73	79	2	6.02	91	99	1	6.68
74	80	2	9.39	92	100	1	10.05
75	82		.76	93	101	2	1.42
76	83		4.13	94	102	2	4.79
77	84		7.50	95	103	2	8.16
78	85		10.87	96	104	2	11.53
79	86	1	2.24	97	106		2.90
80	87	1	5.61	98	107		6.27
81	88	1	8.98	99	108		9.64
82	89	2	.35	100	109	1	1.01

Table 30 Meters to Yards

1 meter = 1.093 yards

m.	yd.	m.	yd.	m.	yd.
1	1.0936	33	36.0893	65	71.0849
2	2.1872	34	37.1829	66	72.1785
3	3.2808	35	38.2765	67	73.2721
4	4.3745	36	39.3701	68	74.3657
5	5.4681	37	40.4637	69	75.4593
6	6.5617	38	41.5573	70	76.5529
7	7.6553	39	42.6509	71	77.6465
8	8.7489	40	43.7445	72	78.7402
9	9.8425	41	44.8381	73	79.8338
10	10.9361	42	45.9318	74	80.9274
11	12.0297	43	47.0254	75	82.0210
12	13.1234	44	48.1190	76	83.1146
13	14.2170	45	49.2126	77	84.2082
14	15.3106	46	50.3062	78	85.3018
15	16.4042	47	51.3998	79	86.3955
16	17.4978	48	52.4934	80	87.4891
17	18.5914	49	53.5871	81	88.5827
18	19.6850	50	54.6807	82	89.6763
19	20.7787	51	55.7743	83	90.7699
20	21.8723	52	56.8679	84	91.8635
21	22.9659	53	57.9615	85	92.9571
22	24.0595	54	59.0551	86	94.0507
23	25.1531	55	60.1487	87	95.1444
24	26.2467	56	61.2423	88	96.2380
25	27.3403	57	62.3360	89	97.3316
26	28.4339	58	63.4296	90	98.4252
27	29.5276	59	64.5232	91	99.5188
28	30.6212	60	65.6168	92	100.6124
29	31.7148	61	66.7104	93	101.7060
30	32.8084	62	67.8040	94	102.7997
31	33.9020	63	68.8976	95	103.8933
32	34.9956	64	69.9913	96	104.9869

Table 30 *(continued)*

m.	yd.	m.	yd.	m.	yd.
97	106.0805	131	143.2633	165	180.4462
98	107.1741	132	144.3570	166	181.5398
99	108.2677	133	145.4506	167	182.6334
100	109.3613	134	146.5442	168	183.7270
101	110.4549	135	147.6378	169	184.8206
102	111.5486	136	148.7314	170	185.9143
103	112.6422	137	149.8250	171	187.0079
104	113.7358	138	150.9186	172	188.1015
105	114.8294	139	152.0122	173	189.1951
106	115.9230	140	153.1059	174	190.2887
107	117.0166	141	154.1995	175	191.3823
108	118.1102	142	155.2931	176	192.4759
109	119.2038	143	156.3867	177	193.5696
110	120.2975	144	157.4803	178	194.6632
111	121.3911	145	158.5739	179	195.7568
112	122.4847	146	159.6675	180	196.8504
113	123.5783	147	160.7612	181	197.9440
114	124.6719	148	161.8548	182	199.0376
115	125.7655	149	162.9484	183	200.1312
116	126.8591	150	164.0420	184	201.2248
117	127.9528	151	165.1356	185	202.3185
118	129.0464	152	166.2292	186	203.4121
119	130.1400	153	167.3228	187	204.5057
120	131.2336	154	168.4164	188	205.5993
121	132.3272	155	169.5101	189	206.6929
122	133.4208	156	170.6037	190	207.7865
123	134.5144	157	171.6973	191	208.8801
124	135.6080	158	172.7909	192	209.9738
125	136.7017	159	173.8845	193	211.0674
126	137.7953	160	174.9781	194	212.1610
127	138.8889	161	176.0717	195	213.2546
128	139.9825	162	177.1654	196	214.3482
129	141.0761	163	178.2590	197	215.4418
130	142.1697	164	179.3526	198	216.5354

Table 30 *(continued)*

m.	yd.	m.	yd.	m.	yd.
199	217.6290	233	254.8119	267	291.9947
200	218.7227	234	255.9055	268	293.0884
201	219.8163	235	256.9991	269	294.1820
202	220.9099	236	258.0927	270	295.2756
203	222.0035	237	259.1863	271	296.3692
204	223.0971	238	260.2800	272	297.4628
205	224.1907	239	261.3736	273	298.5564
206	225.2843	240	262.4672	274	299.6500
207	226.3780	241	263.5608	275	300.7437
208	227.4716	242	264.6544	276	301.8373
209	228.5652	243	265.7480	277	302.9309
210	229.6588	244	266.8416	278	304.0245
211	230.7524	245	267.9353	279	305.1181
212	231.8460	246	269.0289	280	306.2117
213	232.9396	247	270.1225	281	307.3053
214	234.0332	248	271.2161	282	308.3989
215	235.1269	249	272.3097	283	309.4926
216	236.2205	250	273.4033	284	310.5862
217	237.3141	251	274.4969	285	311.6798
218	238.4077	252	275.5905	286	312.7734
219	239.5013	253	276.6842	287	313.8670
220	240.5949	254	277.7778	288	314.9606
221	241.6885	255	278.8714	289	316.0542
222	242.7822	256	279.9650	290	317.1479
223	243.8758	257	281.0586	291	318.2415
224	244.9694	258	282.1522	292	319.3351
225	246.0630	259	283.2458	293	320.4287
226	247.1566	260	284.3395	294	321.5223
227	248.2502	261	285.4331	295	322.6159
228	249.3438	262	286.5267	296	323.7095
229	250.4374	263	287.6203	297	324.8031
230	251.5311	264	288.7139	298	325.8968
231	252.6247	265	289.8075	299	326.9904
232	253.7183	266	290.9011	300	328.0840

Table 30 *(continued)*

m.	yd.	m.	yd.	m.	yd.
301	329.1776	335	366.3605	369	403.5433
302	330.2712	336	367.4541	370	404.6369
303	331.3648	337	368.5477	371	405.7305
304	332.4584	338	369.6413	372	406.8241
305	333.5521	339	370.7349	373	407.9178
306	334.6457	340	371.8285	374	409.0114
307	335.7393	341	372.9221	375	410.1050
308	336.8329	342	374.0157	376	411.1986
309	337.9265	343	375.1094	377	412.2922
310	339.0201	344	376.2030	378	413.3858
311	340.1137	345	377.2966	379	414.4794
312	341.2073	346	378.3902	380	415.5731
313	342.3010	347	379.4838	381	416.6667
314	343.3946	348	380.5774	382	417.7603
315	344.4882	349	381.6710	383	418.8539
316	345.5818	350	382.7647	384	419.9475
317	346.6754	351	383.8583	385	421.0411
318	347.7690	352	384.9519	386	422.1347
319	348.8626	353	386.0455	387	423.2283
320	349.9563	354	387.1391	388	424.3220
321	351.0499	355	388.2327	389	425.4156
322	352.1435	356	389.3263	390	426.5092
323	353.2371	357	390.4199	391	427.6028
324	354.3307	358	391.5136	392	428.6964
325	355.4243	359	392.6072	393	429.7900
326	356.5179	360	393.7008	394	430.8836
327	357.6115	361	394.7944	395	431.9772
328	358.7052	362	395.8880	396	433.0709
329	359.7988	363	396.9816	397	434.1645
330	360.8924	364	398.0752	398	435.2581
331	361.9860	365	399.1688	399	436.3517
332	363.0796	366	400.2625	400	437.4453
333	364.1732	367	401.3561	401	438.5389
334	365.2668	368	402.4497	402	439.6325

Table 30 *(continued)*

m.	yd.	m.	yd.	m.	yd.
403	440.7262	437	477.9090	471	515.0919
404	441.8198	438	479.0026	472	516.1855
405	442.9134	439	480.0962	473	517.2791
406	444.0070	440	481.1898	474	518.3727
407	445.1006	441	482.2835	475	519.4663
408	446.1942	442	483.3711	476	520.5599
409	447.2878	443	484.4707	477	521.6523
410	448.3815	444	485.5643	478	522.7472
411	449.4751	445	486.6579	479	523.8408
412	450.5687	446	487.7515	480	524.9344
413	451.6623	447	488.8451	481	526.0280
414	452.7559	448	489.9388	482	572.1216
415	453.8495	449	491.0324	483	528.2152
416	454.9431	450	492.1260	484	529.3088
417	456.0367	451	493.2196	485	530.4024
418	457.1304	452	494.3132	486	531.4961
419	458.2240	453	495.4068	487	532.5897
420	459.3176	454	496.5004	488	533.6833
421	460.4112	455	497.5940	489	534.7769
422	461.5048	456	498.6877	490	535.8705
423	462.5984	457	499.7813	491	536.9641
424	463.6920	458	500.8749	492	538.0577
425	464.7856	459	501.9685	493	539.1514
426	465.8793	460	503.0621	494	540.2450
427	466.9729	461	504.1557	495	541.3386
428	468.0665	462	505.2493	496	542.4522
429	469.1601	463	506.3430	497	543.5258
430	470.2537	464	507.4366	498	544.6194
431	471.3473	465	508.5302	499	545.7130
432	472.4409	466	509.6238	500	546.8066
433	473.5346	467	510.7174	510	557.7428
434	474.6282	468	511.8110	520	568.6789
435	475.7218	469	512.9046	530	579.6150
436	476.8154	470	513.9982	540	590.5512

Table 30 *(continued)*

m.	yd.	m.	yd.	m.	yd.
550	601.4873	710	776.4654	860	940.5074
560	612.4234	720	787.4016	870	951.4436
570	623.3596	730	798.3377	880	962.3797
580	634.2957	740	809.2738	890	973.3158
590	645.2318	750	820.2100	900	984.2520
600	656.1680	760	831.1461	910	995.1881
610	667.1041	770	842.0822	920	1006.1242
620	678.0402	780	853.0184	930	1017.0604
630	688.9764	790	863.9545	940	1027.9965
640	699.9125	800	874.8906	950	1038.9326
650	710.8486	810	885.8268	960	1049.8688
660	721.7848	820	896.7629	970	1060.8049
670	732.7209	830	907.6990	980	1071.7410
680	743.6570	840	918.6352	990	1082.6772
690	754.5932	850	929.5713	1000	1093.6133
700	765.5293				

Table 31 Kilometers to Miles

1 kilometer = .6214 mile

km.	mi.	km.	mi.	km.	mi.
1	.6214	33	20.5052	65	40.3891
2	1.2427	34	21.1266	66	41.0105
3	1.8641	35	21.7480	67	41.6319
4	2.4855	36	22.3694	68	42.2532
5	3.1069	37	22.9907	69	42.8746
6	3.7282	38	23.6121	70	43.4960
7	4.3496	39	24.2335	71	44.1174
8	4.9710	40	24.8548	72	44.7387
9	5.5923	41	25.4762	73	45.3601
10	6.2137	42	26.0976	74	45.9815
11	6.8351	43	26.7190	75	46.6028
12	7.4565	44	27.3403	76	47.2242
13	8.0778	45	27.9617	77	47.8456
14	8.6992	46	28.5831	78	48.4670
15	9.3206	47	29.2044	79	49.0883
16	9.9419	48	29.8258	80	49.7097
17	10.5633	49	30.4472	81	50.3311
18	11.1847	50	31.0686	82	50.9524
19	11.8061	51	31.6899	83	51.5738
20	12.4274	52	32.3113	84	52.1952
21	13.0488	53	32.9327	85	52.8166
22	13.6702	54	33.5540	86	53.4379
23	14.2915	55	34.1754	87	54.9593
24	14.9129	56	34.7968	88	54.6807
25	15.5343	57	35.4182	89	55.3020
26	16.1557	58	36.0395	90	55.9234
27	16.7770	59	36.6609	91	56.5448
28	17.3984	60	37.2823	92	57.1661
29	18.0198	61	37.9036	93	57.7875
30	18.6411	62	38.5250	94	58.4089
31	19.2625	63	39.1464	95	59.0303
32	19.8839	64	39.7678	96	59.6516

Table 31 *(continued)*

km.	mi.	km.	mi.	km.	mi.
97	60.2730	131	81.3996	165	102.5262
98	60.8944	132	82.0210	166	103.1476
99	61.5157	133	82.6424	167	103.7690
100	62.1371	134	83.2637	168	104.3904
101	62.7585	135	83.8851	169	105.0117
102	63.3799	136	84.5065	170	105.6331
103	64.0012	137	85.1279	171	106.2545
104	64.6226	138	85.7492	172	106.8758
105	65.2440	139	86.3706	173	107.4972
106	65.8653	140	86.9920	174	108.1186
107	66.4867	141	87.6133	175	108.7400
108	67.1081	142	88.2347	176	109.3613
109	67.7295	143	88.8561	177	109.9827
110	68.3508	144	89.4775	178	110.6041
111	68.9722	145	90.0988	179	111.2254
112	69.5936	146	90.7202	180	111.8468
113	70.2149	147	91.3416	181	112.4682
114	70.8363	148	91.9629	182	113.0896
115	71.4577	149	92.5843	183	113.7109
116	72.0791	150	93.2057	184	114.3323
117	72.7004	151	93.8270	185	114.9537
118	73.3218	152	94.4484	186	115.5750
119	73.9432	153	95.0698	187	116.1964
120	74.5645	154	95.6912	188	116.8178
121	75.1859	155	96.3125	189	117.4392
122	75.8073	156	96.9339	190	118.0605
123	76.4287	157	97.5553	191	118.6819
124	77.0500	158	98.1766	192	119.3033
125	77.6714	159	98.7980	193	119.9246
126	78.2928	160	99.4194	194	120.5460
127	78.9141	161	100.0408	195	121.1674
128	79.5355	162	100.6621	196	121.7888
129	80.1569	163	101.2835	197	122.4101
130	80.7783	164	101.9049	198	123.0315

Table 31 *(continued)*

km.	mi.	km.	mi.	km.	mi.
199	123.6529	233	144.7795	267	165.9061
200	124.2742	234	145.4009	268	166.5275
201	124.8956	235	146.0222	269	167.1489
202	125.5170	236	146.6436	270	167.7702
203	126.1384	237	147.2650	271	168.3916
204	126.7597	238	147.8863	272	169.0130
205	127.3811	239	148.5077	273	169.6343
206	128.0025	240	149.1291	274	170.2557
207	128.6238	241	149.7505	275	170.8771
208	129.2452	242	150.3718	276	171.4984
209	129.8666	243	150.9932	277	172.1198
210	130.4880	244	151.6146	278	172.7412
211	131.1093	245	152.2359	279	173.3626
212	131.7307	246	152.8573	280	173.9839
213	132.3521	247	153.4787	281	174.6053
214	132.9734	248	154.1001	282	175.2267
215	133.5948	249	154.7214	283	175.8483
216	134.2162	250	155.3428	284	176.4694
217	134.8375	251	155.9642	285	177.0908
218	135.4589	252	156.5855	286	177.7122
219	136.0803	253	157.2069	287	178.3335
220	136.7017	254	157.8283	288	178.9549
221	137.3230	255	158.4497	289	179.5763
222	137.9444	256	159.0710	290	180.1976
223	138.5658	257	159.6924	291	180.8190
224	139.1871	258	160.3138	292	181.4404
225	139.8085	259	160.9351	293	182.0618
226	140.4299	260	161.5565	294	182.6831
227	141.0513	261	162.1779	295	183.3045
228	141.6726	262	162.7993	296	183.9259
229	142.2940	263	163.4206	297	184.5472
230	142.9154	264	164.0420	298	185.1686
231	143.5367	265	164.6634	299	185.7900
232	144.1581	266	165.2847	300	186.4114

Table 31 *(continued)*

km.	mi.	km.	mi.	km.	mi.
301	187.0327	335	208.1593	369	229.2860
302	187.6541	336	208.7807	370	229.9073
303	188.2755	337	209.4021	371	230.5287
304	188.8978	338	210.0235	372	231.1501
305	189.5182	339	210.6448	373	231.7715
306	190.1396	340	211.2662	374	232.3928
307	190.7610	341	211.8876	375	233.0142
308	191.3823	342	212.5089	376	233.6356
309	192.0037	343	213.1303	377	234.2569
310	192.6251	344	213.7517	378	234.8783
311	193.2464	345	214.3731	379	235.4997
312	193.8678	346	214.9944	380	236.1211
313	194.4892	347	215.6158	381	236.7424
314	195.1106	348	216.2372	382	237.3638
315	195.7319	349	216.8585	383	237.9852
316	196.3533	350	217.4799	384	238.6065
317	196.9747	351	218.1013	385	239.2279
318	197.5960	352	218.7227	386	239.8493
319	198.2174	353	219.3440	387	240.4707
320	198.8388	354	219.9654	388	241.0920
321	199.4602	355	220.5868	389	241.7134
322	200.0815	356	221.2081	390	242.3348
323	200.7029	357	221.8295	391	242.9561
324	201.3243	358	222.4509	392	243.5775
325	201.9456	359	223.0723	393	244.1989
326	202.5670	360	223.6936	394	244.8202
327	203.1884	361	224.3150	395	245.4416
328	203.8098	362	224.9364	396	246.0630
329	204.4311	363	225.5577	397	246.6844
330	205.0525	364	226.1791	398	247.3057
331	205.6739	365	226.8005	399	247.9271
332	206.2952	366	227.4219	400	248.5485
333	206.9166	367	228.0432	401	249.1698
334	207.5380	368	228.6646	402	249.7912

Table 31 *(continued)*

km.	mi.	km.	mi.	km.	mi.
403	250.4126	437	271.5392	471	292.6658
404	251.0340	238	272.1606	472	293.2872
405	251.6553	439	272.7820	473	293.9086
406	252.2767	440	273.4033	474	294.5299
407	252.8981	441	274.0247	475	295.1513
408	253.5194	442	274.6461	476	295.7727
409	254.1408	443	275.2674	477	296.3941
410	254.7622	444	275.8888	478	297.0154
411	255.3836	445	276.5102	479	297.6368
412	256.0049	446	277.1316	480	298.2582
413	256.6263	447	277.7529	481	298.8795
414	257.2477	448	278.3743	482	299.5009
415	257.8690	449	278.9957	483	300.1223
416	258.4904	450	279.6175	484	300.7437
417	259.1118	451	280.2384	485	301.3650
418	259.7332	452	280.8598	486	301.9864
419	260.3545	453	281.4811	487	302.6078
420	260.9759	454	282.1025	488	303.2291
421	261.5973	455	282.7239	489	303.8505
422	262.2186	456	283.3453	490	304.4719
423	262.8400	457	283.9666	491	305.0933
424	263.4614	458	284.5880	492	305.7146
425	264.0828	459	285.2094	493	306.3360
426	264.7041	460	285.8307	494	306.9574
427	265.3255	461	286.4521	495	307.5787
428	265.9469	462	287.0735	496	308.2001
429	266.5682	463	287.6949	497	308.8215
430	267.1896	464	288.3162	498	309.4429
431	267.8110	465	288.9376	499	310.0642
432	268.4324	466	289.5590	500	310.6856
433	269.0537	467	290.1803	510	316.8993
434	269.6751	468	290.8017	520	323.1130
435	270.2965	469	291.4231	530	329.3267
436	270.9178	470	292.0445	540	335.5404

Table 31 *(continued)*

km.	mi.	km.	mi.	km.	mi.
550	341.7542	760	472.2421	970	602.1087
560	347.9679	770	478.4558	980	608.9438
570	354.1816	780	484.6695	990	615.1575
580	360.3953	790	490.8832	1000	621.3712
590	366.6090	800	497.0970	2000	1242.742
600	372.8227	810	503.3107	3000	1864.114
610	379.9364	820	509.5244	4000	2485.485
620	385.2501	830	515.7381	5000	3106.856
630	391.4639	840	521.9518	6000	3728.227
640	397.6776	850	528.1655	7000	4349.598
650	403.8913	860	534.3792	8000	4970.970
660	410.1050	870	540.5929	9000	5592.341
670	416.3187	880	546.8067	10000	6213.712
680	422.5324	890	553.0204	15000	9320.568
690	428.7461	900	559.2341	20000	12427.42
700	434.9598	910	565.4478	25000	15534.28
710	441.1736	920	571.6615	30000	18641.14
720	447.3873	930	577.8752	35000	21747.99
730	453.6010	940	584.0889	40000	24854.85
740	459.8147	950	590.3026	45000	27961.70
750	466.0284	960	596.5163	50000	31068.56

Table 32 Kilometers to Knots (International Nautical Miles)

1 kilometer = .54 international nautical mile

km.	INM	km.	INM	km.	INM
1	.5400	32	17.2786	63	34.0173
2	1.0799	33	17.8186	64	34.5572
3	1.6199	34	18.3585	65	35.0972
4	2.1598	35	18.8985	66	35.6371
5	2.6998	36	19.4384	67	36.1771
6	3.2397	37	19.9784	68	36.7171
7	3.7797	38	20.5184	69	37.2570
8	4.3197	39	21.0583	70	37.7970
9	4.8596	40	21.5983	71	38.3369
10	5.3996	41	22.1382	72	38.8769
11	5.9395	42	22.6782	73	39.4168
12	6.4795	43	23.2181	74	39.9568
13	7.0194	44	23.7581	75	40.4968
14	7.5594	45	24.2981	76	41.0367
15	8.0994	46	24.8380	77	41.5767
16	8.6393	47	25.3780	78	42.1166
17	9.1793	48	25.9179	79	42.6566
18	9.7192	49	26.4579	80	43.1965
19	10.2592	50	26.9978	81	43.7365
20	10.7991	51	27.5378	82	44.2765
21	11.3391	52	28.0778	83	44.8164
22	11.8790	53	28.6177	84	45.3564
23	12.4190	54	29.1577	85	45.8963
24	12.9590	55	29.6976	86	46.4363
25	13.4989	56	30.2376	87	46.9762
26	14.0389	57	30.7775	88	47.5162
27	14.5788	58	31.3175	89	48.0562
28	15.1188	59	31.8575	90	48.5961
29	15.6587	60	32.3974	91	49.1361
30	16.1987	61	32.9374	92	49.6760
31	16.7387	62	33.4773	93	50.2160

Table 32 *(continued)*

km.	INM	km.	INM	km.	INM
94	50.7559	128	69.1145	162	87.4730
95	51.2959	129	69.6544	163	88.0130
96	51.8359	130	70.1944	164	88.5529
97	52.3758	131	70.7343	165	89.0929
98	52.9158	132	71.2743	166	89.6328
99	53.4557	133	71.8143	167	90.1728
100	53.9957	134	72.3542	168	90.7127
101	54.5356	135	72.8942	169	91.2527
102	55.0756	136	73.4341	170	91.7927
103	55.6156	137	73.9741	171	92.3326
104	56.1555	138	74.5140	172	92.8726
105	56.6955	139	75.0540	173	93.4125
106	57.2354	140	75.5940	174	93.9525
107	57.7754	141	76.1339	175	94.4924
108	58.3153	142	76.6739	176	95.0324
109	58.8553	143	77.2138	177	95.5724
110	59.3952	144	77.7538	178	96.1123
111	59.9352	145	78.2937	179	96.6523
112	60.4752	146	78.8337	180	97.1922
113	61.0151	147	79.3736	181	97.7322
114	61.5551	148	79.9136	182	98.2721
115	62.0950	149	80.4536	183	98.8121
116	62.6350	150	80.9935	184	99.3521
117	63.1749	151	81.5335	185	99.8920
118	63.7149	152	82.0734	186	100.4320
119	64.2549	153	82.6134	187	100.9719
120	64.7948	154	83.1533	188	101.5119
121	65.3348	155	83.6933	189	102.0518
122	65.8747	156	84.2333	190	102.5918
123	66.4147	157	84.7732	191	103.1317
124	66.9546	158	85.3132	192	103.6717
125	67.4946	159	85.8531	193	104.2117
126	68.0346	160	86.3931	194	104.7516
127	68.5745	161	86.9330	195	105.2916

Table 32 *(continued)*

km.	INM	km.	INM	km.	INM
196	105.8315	350	188.9849	900	485.9611
197	106.3715	360	194.3844	1000	539.9568
198	106.9114	370	199.7840	2000	1079.914
199	107.4514	380	205.1836	3000	1619.870
200	107.9914	390	210.5832	4000	2159.827
210	113.3909	400	215.9827	5000	2699.784
220	118.7904	410	221.3823	6000	3239.741
230	124.1901	420	226.7819	7000	3779.698
240	129.5896	430	232.1814	8000	4319.654
250	134.9892	440	237.5810	9000	4859.611
260	140.3888	450	242.9806	10000	5399.568
270	145.7883	460	248.3801	15000	8099.352
280	151.1879	470	253.7797	20000	10799.14
290	156.5875	480	259.1793	25000	13498.92
300	161.9870	490	264.5788	30000	16198.70
310	167.3866	500	269.9784	35000	18898.49
320	172.2462	600	323.9741	40000	21598.27
330	178.1857	700	377.9698	45000	24298.06
340	183.5853	800	431.9654	50000	26997.84

Table 33 Kilometers per Liter to Miles per Gallon

1 kilometer per liter = 2.35 miles per gallon

km./l.	mi./gal	km./l.	mi./gal	km./l.	mi./gal
1	2.35	18	42.3	35	82.3
2	4.70	19	44.7	36	84.7
3	7.06	20	47.0	37	87.0
4	9.41	21	49.4	38	89.4
5	11.8	22	51.7	39	91.7
6	14.1	23	54.1	40	94.1
7	16.5	24	56.5	41	96.4
8	18.8	25	58.8	42	98.8
9	21.2	26	61.2	43	101.1
10	23.5	27	63.5	44	103.5
11	25.9	28	65.9	45	105.8
12	28.2	29	68.2	46	108.2
13	30.6	30	70.6	47	110.6
14	32.9	31	72.9	48	112.9
15	35.3	32	75.3	49	115.3
16	37.6	33	77.6	50	117.6
17	40.0	34	80.0		

AREA

Table 34 Square Centimeters to Square Inches

1 square centimeter = 0.155 square inch

cm.^{2}	in.^{2}	cm.^{2}	in.^{2}	cm.^{2}	in.^{2}
1	0.15500	33	5.11501	65	10.07502
2	0.31000	34	5.27001	66	10.23002
3	0.46500	35	5.42501	67	10.38502
4	0.62000	36	5.58001	68	10.54002
5	0.77500	37	5.73501	69	10.69502
6	0.93000	38	5.89001	70	10.85002
7	1.08500	39	6.04501	71	11.00502
8	1.24000	40	6.20001	72	11.16002
9	1.39500	41	6.35501	73	11.31502
10	1.55000	42	6.51001	74	11.47002
11	1.70500	43	6.66501	75	11.62502
12	1.86000	44	6.82001	76	11.78002
13	2.01500	45	6.97501	77	11.93502
14	2.17000	46	7.13001	78	12.09002
15	2.32500	47	7.28501	79	12.24502
16	2.48000	48	7.44001	80	12.40002
17	2.63501	49	7.59502	81	12.55503
18	2.79001	50	7.75002	82	12.71003
19	2.94501	51	7.90502	83	12.86503
20	3.10001	52	8.06002	84	13.02003
21	3.25501	53	8.21502	85	13.17503
22	3.41001	54	8.37002	86	13.33003
23	3.56501	55	8.52502	87	13.48503
24	3.72001	56	8.68002	88	13.64003
25	3.87501	57	8.83502	89	13.79503
26	4.03001	58	8.99002	90	13.95003
27	4.18501	59	9.14502	91	14.10503
28	4.34001	60	9.30002	92	14.26003
29	4.49501	61	9.45502	93	14.41503
30	4.65001	62	9.61002	94	14.57003
31	4.80501	63	9.76502	95	14.72503
32	4.96001	64	9.92002	96	14.88003

Table 34 *(continued)*

cm.^{2}	in.^{2}	cm.^{2}	in.^{2}	cm.^{2}	in.^{2}
97	15.03503	131	20.30504	165	25.57505
98	15.19003	132	20.46004	166	25.73005
99	15.34503	133	20.61504	167	25.88505
100	15.50003	134	20.77004	168	26.04005
101	15.65503	135	20.92504	169	26.19505
102	15.81003	136	21.08004	170	26.35005
103	15.96503	137	21.23504	171	26.50505
104	16.12003	138	21.39004	172	26.66005
105	16.27503	139	21.54504	173	26.81505
106	16.43003	140	21.70004	174	26.97005
107	16.58503	141	21.85504	175	27.12505
108	17.74003	142	22.01004	176	27.28005
109	16.89503	143	22.16504	177	27.43505
110	17.05003	144	22.32004	178	27.59006
111	17.20503	145	22.47504	179	27.74506
112	17.36003	146	22.63005	180	27.90006
113	17.51503	147	22.78505	181	28.05506
114	17.67004	148	22.94005	182	28.21006
115	17.82504	149	23.09505	183	28.36506
116	17.98004	150	23.25005	184	28.52006
117	18.13504	151	23.40505	185	28.67506
118	18.29004	152	23.56005	186	28.83006
119	18.44504	153	23.71505	187	28.98506
120	18.60004	154	23.87005	188	29.14006
121	18.75504	155	24.02505	189	29.29506
122	18.91004	156	24.18005	190	29.45006
123	19.06504	157	24.33505	191	29.60506
124	19.22004	158	24.49005	192	29.76006
125	19.37504	159	24.64505	193	29.91506
126	19.53004	160	24.80005	194	30.07006
127	19.68504	161	24.95505	195	30.22506
128	19.84004	162	25.11005	196	30.38006
129	19.99504	163	25.26505	197	30.53506
130	20.15004	164	25.42005	198	30.69006

Table 34 *(continued)*

cm.^{2}	in.^{2}	cm.^{2}	in.^{2}	cm.^{2}	in.^{2}
199	30.84506	350	54.25011	500	77.50015
200	31.00006	360	55.80011	600	93.00019
210	32.55007	370	57.35011	700	108.500
220	34.10007	380	58.90012	800	124.000
230	35.65007	390	60.45012	900	139.500
240	37.20007	400	62.00012	1000	155.00
250	38.75008	410	63.55013	2000	310.00
260	40.30008	420	65.10013	3000	465.00
270	41.85008	430	66.65013	4000	620.00
280	43.40009	440	68.20014	5000	775.00
290	44.95009	450	69.75014	6000	930.00
300	46.50009	460	71.30014	7000	1085.0
310	48.05010	470	72.85015	8000	1240.0
320	49.60010	480	74.40015	9000	1395.0
330	51.15010	490	75.95015	10000	1550.0
340	52.70011				

Table 35 Square Meters to Square Feet

1 square meter = 10.764 square feet

$m.^2$	$ft.^2$	$m.^2$	$ft.^2$	$m.^2$	$ft.^2$
1	10.7639	33	355.2090	65	699.6542
2	21.5278	34	365.9730	66	710.4181
3	32.2917	35	376.7369	67	721.1820
4	43.0556	36	387.5008	68	731.9459
5	53.8196	37	398.2647	69	742.7098
6	64.5835	38	409.0286	70	753.4737
7	75.3474	39	419.7925	71	764.2376
8	86.1113	40	430.5564	72	775.0015
9	96.8752	41	441.3203	73	785.7655
10	107.6391	42	452.0842	74	796.5294
11	118.4030	43	462.8481	75	807.2933
12	129.1669	44	473.6121	76	818.0572
13	139.9308	45	484.3760	77	828.8211
14	150.6947	46	495.1399	78	839.5850
15	161.4587	47	505.9038	79	850.3489
16	172.2226	48	516.6677	80	861.1128
17	182.9865	49	527.4316	81	871.8767
18	193.7504	50	538.1955	82	882.6407
19	204.5143	51	548.9594	83	893.4046
20	215.2782	52	559.7233	84	904.1685
21	226.0421	53	570.4873	85	914.9324
22	236.8060	54	581.2512	86	925.6963
23	247.5699	55	592.0151	87	936.4602
24	258.3338	56	602.7790	88	947.2241
25	269.0978	57	613.5429	89	957.9880
26	279.8617	58	624.3068	90	968.7519
27	290.6256	59	635.0707	91	979.5158
28	301.3895	60	645.8346	92	990.2798
29	312.1534	61	656.5985	93	1001.044
30	322.9173	62	667.3624	94	1011.808
31	333.6812	63	678.1264	95	1022.571
32	344.4451	64	688.8903	96	1033.335

Table 35 *(continued)*

m.^{2}	ft.^{2}	m.^{2}	ft.^{2}	m.^{2}	ft.^{2}
97	1044.099	131	1410.072	165	1776.045
98	1054.863	132	1420.836	166	1786.809
99	1065.627	133	1431.600	167	1797.573
100	1076.391	134	1442.364	168	1808.337
101	1087.155	135	1453.128	169	1819.101
102	1097.919	136	1463.892	170	1829.865
103	1108.683	137	1474.656	171	1840.629
104	1119.447	138	1485.420	172	1851.393
105	1130.211	139	1496.184	173	1862.156
106	1140.975	140	1506.947	174	1872.920
107	1151.738	141	1517.711	175	1883.684
108	1162.502	142	1528.475	176	1894.448
109	1173.266	143	1539.239	177	1905.212
110	1184.030	144	1550.003	178	1915.976
111	1194.794	145	1560.767	179	1926.740
112	1205.558	146	1571.531	180	1937.504
113	1216.322	147	1582.295	181	1948.268
114	1227.086	148	1593.059	182	1959.032
115	1237.850	149	1603.823	183	1969.796
116	1248.614	150	1614.587	184	1980.560
117	1259.378	151	1625.350	185	1991.323
118	1270.141	152	1636.114	186	2002.087
119	1280.905	153	1646.878	187	2012.851
120	1291.669	154	1657.642	188	2023.615
121	1302.433	155	1668.406	189	2034.379
122	1313.197	156	1679.170	190	2045.143
123	1323.961	157	1689.934	191	2055.907
124	1334.725	158	1700.698	192	2066.671
125	1345.489	159	1711.462	193	2077.435
126	1356.253	160	1722.226	194	2088.199
127	1367.017	161	1732.990	195	2098.963
128	1377.781	162	1743.753	196	2109.726
129	1388.544	163	1754.517	197	2120.490
130	1399.308	164	1765.281	198	2131.254

Table 35 *(continued)*

$m.^2$	$ft.^2$	$m.^2$	$ft.^2$	$m.^2$	$ft.^2$
199	2142.018	800	8611.128	5000	53819.55
200	2152.782	900	1687.519	6000	64583.46
300	3229.173	1000	10763.91	7000	75347.37
400	4305.564	2000	21527.82	8000	86111.28
500	5381.955	3000	32291.73	9000	96875.19
600	6458.346	4000	43055.64	10000	107639.1
700	7534.737				

Table 36 Square Meters to Square Yards

1 square meter = 1.196 square yards

m.²	yd.²	m.²	yd.²	m.²	yd.²
1	1.1960	33	39.4677	65	77.7394
2	2.3920	34	40.6637	66	78.9353
3	3.5880	35	41.8597	67	80.1313
4	4.7840	36	43.0556	68	81.3273
5	5.9800	37	44.2516	69	82.5233
6	7.1759	38	45.4476	70	83.7193
7	8.3719	39	46.6436	71	84.9153
8	9.5679	40	47.8396	72	86.1113
9	10.7639	41	49.0356	73	87.3073
10	11.9599	42	50.2316	74	88.5033
11	13.1559	43	51.4276	75	89.6993
12	14.3519	44	52.6236	76	90.8952
13	15.5479	45	53.8196	77	92.0912
14	16.7439	46	55.0155	78	93.2872
15	17.9399	47	56.2115	79	94.4832
16	19.1358	48	57.4075	80	95.6792
17	20.3318	49	58.6035	81	96.8752
18	21.5278	50	59.7995	82	98.0712
19	22.7238	51	60.9955	83	99.2672
20	23.9198	52	62.1915	84	100.4632
21	25.1158	53	63.3875	85	101.6592
22	26.3118	54	64.5835	86	102.8551
23	27.5078	55	65.7795	87	104.0511
24	28.7038	56	66.9754	88	105.2471
25	29.8998	57	68.1714	89	106.4431
26	31.0957	58	69.3674	90	107.6391
27	32.2917	59	70.5634	91	108.8351
28	33.4877	60	71.7594	92	110.0311
29	34.6837	61	72.9554	93	111.2271
30	35.8797	62	74.1514	94	112.4231
31	37.0757	63	75.3474	95	113.6191
32	38.2717	64	76.5434	96	114.8150

Table 36 *(continued)*

m.^{2}	yd.^{2}	m.^{2}	yd.^{2}	m.^{2}	yd.^{2}
97	116.0110	131	156.6747	165	197.3384
98	117.2070	132	157.8707	166	198.5343
99	118.4030	133	159.0667	167	199.7303
100	119.5990	134	160.2627	168	200.9263
101	120.7950	135	161.4587	169	202.1223
102	121.9910	136	162.6546	170	203.3183
103	123.1870	137	163.8506	171	204.5143
104	124.3830	138	165.0466	172	205.7103
105	125.5790	139	166.2426	173	206.9063
106	126.7749	140	167.4386	174	208.1023
107	127.9709	141	168.6346	175	209.2983
108	129.1669	142	169.8306	176	210.4942
109	130.3629	143	171.0266	177	211.6902
110	131.5589	144	172.2226	178	212.8862
111	132.7549	145	173.4186	179	214.0822
112	133.9509	146	174.6145	180	215.2782
113	135.1469	147	175.8105	181	216.4742
114	136.3429	148	177.0065	182	217.6702
115	137.5389	149	178.2025	183	218.8662
116	138.7348	150	179.3985	184	220.0622
117	139.9308	151	180.5945	185	221.2582
118	141.1268	152	181.7905	186	222.4541
119	142.3228	153	182.9865	187	223.6501
120	143.5188	154	184.1825	188	224.8461
121	144.7148	155	185.3785	189	226.0421
122	145.9108	156	186.5744	190	227.2381
123	147.1068	157	187.7704	191	228.4341
124	148.3028	158	188.9664	192	229.6301
125	149.4988	159	190.1624	193	230.8261
126	150.6947	160	191.3584	194	232.0221
127	151.8907	161	192.5544	195	233.2181
128	153.0867	162	193.7504	196	234.4140
129	154.2827	163	194.9464	197	235.6100
130	155.4787	164	196.1424	198	236.8060

Table 36 *(continued)*

m.^{2}	yd.^{2}	m.^{2}	yd.^{2}	m.^{2}	yd.^{2}
199	238.0020	233	278.6657	267	319.3293
200	239.1980	234	279.8617	268	320.5253
201	240.3940	235	281.0577	269	321.7213
202	241.5900	236	282.2536	270	322.9173
203	242.7860	237	283.4496	271	324.1133
204	243.9820	238	284.6456	272	325.3093
205	245.1780	239	285.8416	273	326.5053
206	246.3739	240	287.0376	274	327.7013
207	247.5699	241	288.2336	275	328.8973
208	248.7659	242	289.4296	276	330.0933
209	249.9619	243	290.6256	277	331.2892
210	251.1579	244	291.8216	278	332.4852
211	252.3539	245	293.0176	279	333.6812
212	253.5499	246	294.2135	280	334.8772
213	254.7459	247	295.4095	281	336.0732
214	255.9419	248	296.6055	282	337.2692
215	257.1379	249	297.8015	283	338.4652
216	258.3338	250	298.9975	284	339.6612
217	259.5298	251	300.1935	285	340.8572
218	260.7258	252	301.3895	286	342.0532
219	261.9218	253	302.5855	287	343.2491
220	263.1178	254	303.7815	288	344.4451
221	264.3138	255	304.9775	289	345.6411
222	265.5098	256	306.1735	290	346.8371
223	266.7058	257	307.3694	291	348.0331
224	267.0978	258	308.5654	292	349.2291
225	267.9018	259	309.7614	293	350.4251
226	270.2937	260	310.9574	294	351.6211
227	271.4897	261	312.1534	295	352.8171
228	272.6857	262	313.3494	296	354.0131
229	273.8817	263	314.5454	297	355.2090
230	275.0777	264	315.7414	298	356.4050
231	276.2737	265	316.9374	299	357.6010
232	277.4697	266	318.1334	300	358.7970

Table 36 *(continued)*

m.^{2}	yd.^{2}	m.^{2}	yd.^{2}	m.^{2}	yd.^{2}
301	359.9930	335	400.6567	369	441.3203
302	361.1890	336	401.8527	370	442.5163
303	362.3850	337	403.0486	371	443.7123
304	363.5810	338	404.2446	372	444.9083
305	364.7770	339	405.4406	373	446.1043
306	365.9729	340	406.6366	374	447.3003
307	367.1689	341	407.8326	375	448.4963
308	368.3649	342	409.0286	376	449.6923
309	369.5609	343	410.2246	377	450.8882
310	370.7569	344	411.4206	378	452.0842
311	371.9529	345	412.6166	379	453.2802
312	373.1489	346	413.8126	380	454.4762
313	374.3449	347	415.0085	381	455.6722
314	375.5409	348	416.2045	382	456.8682
315	376.7369	349	417.4005	383	458.0642
316	377.9328	350	418.5965	384	459.2602
317	379.1288	351	419.7925	385	460.4562
318	380.3248	352	420.9885	386	461.6522
319	381.5208	353	422.1845	387	462.8481
320	382.7168	354	423.3805	388	464.0441
321	383.9128	355	424.5765	389	465.2401
322	385.1088	356	425.7725	390	466.4361
323	386.3048	357	426.9684	391	467.6321
324	387.5008	358	428.1644	392	468.8281
325	388.6968	359	429.3604	393	470.0241
326	389.8927	360	430.5564	394	471.2201
327	391.0887	361	431.7524	395	472.4161
328	392.2847	362	432.9484	396	473.6121
329	393.4807	363	434.1444	397	474.8080
330	394.6767	364	435.3404	398	476.0040
331	395.8727	365	436.5364	399	477.2000
332	397.0687	366	437.7324	400	478.3960
333	398.2647	367	438.9283	401	479.5920
334	399.4607	368	440.1243	402	480.7880

Table 36 *(continued)*

m.^{2}	yd.^{2}	m.^{2}	yd.^{2}	m.^{2}	yd.^{2}
403	481.9840	437	522.6476	471	563.3113
404	483.1800	438	523.8436	472	564.5073
405	484.3760	439	525.0396	473	565.7033
406	485.5720	440	526.2356	474	566.8993
407	486.7679	441	527.4316	475	568.0953
408	487.9639	442	528.6276	476	569.2913
409	489.1599	443	529.8236	477	570.4872
410	490.3559	444	531.0196	478	571.6832
411	491.5519	445	532.2156	479	572.8792
412	492.7479	446	533.4116	480	574.0752
413	493.9439	447	534.6075	481	575.2712
414	495.1399	448	535.8035	482	576.4672
415	496.3359	449	536.9995	483	577.6632
416	497.5319	450	538.1955	484	578.8592
417	498.7278	451	539.3915	485	580.0552
418	499.9238	452	540.5875	486	581.2512
419	501.1198	453	541.7835	487	582.4471
420	502.3158	454	542.9795	488	583.6431
421	503.5118	455	544.1755	489	584.8391
422	504.7078	456	545.3715	490	586.0351
423	505.9038	457	546.5674	491	587.2311
424	507.0998	458	547.7634	492	588.4271
425	508.2958	459	548.9594	493	589.6231
426	509.4918	460	550.1554	494	590.8191
427	510.6877	461	551.3514	495	592.0151
428	511.8837	462	552.5474	496	593.2111
429	513.0797	463	553.7434	497	594.4070
430	514.2757	464	554.9394	498	595.6030
431	515.4717	465	556.1354	499	596.7990
432	516.6677	466	557.3314	500	597.9950
433	517.8637	467	558.5273	510	609.9549
434	519.0597	468	559.7233	520	621.9148
435	520.2557	469	560.9193	530	633.8747
436	521.4517	470	562.1153	540	645.8346

Table 36 *(continued)*

m.^{2}	yd.^{2}	m.^{2}	yd.^{2}	m.^{2}	yd.^{2}
550	657.7945	760	908.9524	970	1160.1103
560	669.7544	770	920.9123	980	1172.0702
570	681.7143	780	932.8722	990	1184.0301
580	693.6742	790	944.8321	1000	1195.99
590	705.6341	800	956.7920	2000	2391.98
600	717.5940	810	968.7519	3000	3587.97
610	729.5539	820	980.7118	4000	4783.96
620	741.5138	830	992.6717	5000	5979.95
630	753.4737	840	1004.6316	6000	7175.94
640	765.4336	850	1016.5915	7000	8371.93
650	777.3935	860	1028.5514	8000	9567.92
660	789.3534	870	1040.5113	9000	10763.91
670	801.3133	880	1052.4712	10000	11959.90
680	813.2732	890	1064.4311	15000	17939.85
690	825.2331	900	1076.3910	20000	23919.80
700	837.1930	910	1088.3509	25000	29899.75
710	849.1529	920	1100.3108	30000	35878.70
720	861.1128	930	1112.2707	35000	41859.65
730	873.0727	940	1124.2306	40000	47839.60
740	885.0326	950	1136.1905	45000	53819.55
750	896.9925	960	1148.1504	50000	59799.50

Table 37 Square Kilometers to Square Miles

1 square kilometer = .3861 square mile

km.^{2}	mi.^{2}	km.^{2}	mi.^{2}	km.^{2}	mi.^{2}
1	.3861	33	12.7414	65	25.0966
2	.7722	34	13.1275	66	25.4827
3	1.1583	35	13.5136	67	25.8688
4	1.5444	36	13.8997	68	26.2549
5	1.9305	37	14.2858	69	26.6410
6	2.3166	38	14.6719	70	27.0272
7	2.7027	39	15.0580	71	27.4133
8	3.0888	40	15.4441	72	27.7994
9	3.4749	41	15.8302	73	28.1855
10	3.8610	42	16.2163	74	28.5716
11	4.2471	43	16.6024	75	28.9577
12	4.6332	44	16.9885	76	29.3438
13	5.0193	45	17.3746	77	29.7299
14	5.4054	46	17.7607	78	30.1160
15	5.7915	47	18.1468	79	30.5021
16	6.1776	48	18.5329	80	30.8882
17	6.5637	49	18.9190	81	31.2743
18	6.9498	50	19.3051	82	31.6604
19	7.3359	51	19.6912	83	32.0465
20	7.7220	52	20.0773	84	32.4326
21	8.1081	53	20.4634	85	32.8187
22	8.4942	54	20.8495	86	33.2048
23	8.8803	55	21.2356	87	33.5909
24	9.2665	56	21.6217	88	33.9770
25	9.6526	57	22.0078	89	34.3631
26	10.0387	58	22.3939	90	34.7492
27	10.4248	59	22.7800	91	35.1353
28	10.8109	60	23.1661	92	35.5214
29	11.1970	61	23.5522	93	35.9075
30	11.5831	62	23.9383	94	36.2936
31	11.9692	63	24.3244	95	36.6797
32	12.3553	64	24.7105	96	37.0658

Table 37 *(continued)*

km.^{2}	mi.^{2}	km.^{2}	mi.^{2}	km.^{2}	mi.^{2}
97	37.4519	410	158.302	750	289.577
98	37.8380	420	162.163	760	293.438
99	38.2241	430	166.024	770	297.299
100	38.6102	440	169.885	780	301.160
110	42.4712	450	173.746	790	305.021
120	46.3323	460	177.607	800	308.882
130	50.1933	470	181.468	810	312.743
140	54.0543	480	185.329	820	316.604
150	57.9153	490	189.190	830	320.465
160	61.7763	500	193.051	840	324.326
170	65.6374	510	196.912	850	328.187
180	69.4984	520	200.773	860	332.048
190	73.3594	530	204.634	870	335.909
200	77.2204	540	208.495	880	339.770
210	81.0815	550	212.356	890	343.631
220	84.9425	560	216.217	900	347.492
230	88.8035	570	220.078	910	351.353
240	92.6645	580	223.939	920	355.214
250	96.5255	590	227.800	930	359.075
260	100.387	600	231.661	940	362.936
270	104.248	610	235.522	950	366.797
280	108.109	620	239.383	960	370.658
290	111.970	630	243.244	970	374.519
300	115.831	640	247.105	980	378.380
310	119.692	650	250.966	990	382.241
320	123.553	660	254.827	1000	386.102
330	127.414	670	258.688	2000	772.20
340	131.275	680	262.549	3000	1158.31
350	135.136	690	266.410	4000	1544.41
360	138.997	700	270.272	5000	1930.51
370	142.858	710	274.133	6000	2316.61
380	146.719	720	277.994	7000	2702.72
390	150.580	730	281.855	8000	3088.82
400	154.441	740	285.716	9000	3474.92

Table 38 Hectares to Acres

1 hectare = 2.471 acres

ha.	acres	ha.	acres	ha.	acres
0.1	.2471	22	54.3632	54	133.4369
0.2	.4942	23	56.8342	55	135.9080
0.25	.6178	24	59.3053	56	138.3790
0.3	.7413	25	61.7763	57	140.8501
0.4	.9884	26	64.2474	58	143.3211
0.5	1.2355	27	66.7185	59	145.7922
0.6	1.4826	28	69.1895	60	148.2632
0.7	1.7297	29	71.6606	61	150.7343
0.75	1.8533	30	74.1316	62	153.2053
0.8	1.9768	31	76.6027	63	155.6764
0.9	2.2239	32	79.0737	64	158.1474
1	2.4711	33	81.5448	65	160.6185
2	4.9421	34	84.0158	66	163.0896
3	7.4132	35	86.4869	67	165.5606
4	9.8842	36	88.9579	68	168.0317
5	12.3553	37	91.4290	69	170.5027
6	14.8263	38	93.9000	70	172.9738
7	17.2874	39	96.3711	71	175.4448
8	19.7684	40	98.8422	72	177.9159
9	22.2395	41	101.3132	73	180.3869
10	24.7105	42	103.7843	74	182.8580
11	27.1816	43	106.2553	75	185.3290
12	29.6526	44	108.7264	76	187.8001
13	32.1237	45	111.1974	77	190.2711
14	34.5948	46	113.6685	78	192.7422
15	37.0658	47	116.1395	79	195.2132
16	39.5369	48	118.6106	80	197.6843
17	42.0079	49	121.0816	81	200.1554
18	44.4790	50	123.5527	82	202.6264
19	46.9500	51	126.0237	83	205.0975
20	49.4211	52	128.4948	84	207.5685
21	51.8921	53	130.9659	85	210.0396

Table 38 *(continued)*

ha.	acres	ha.	acres	ha.	acres
86	212.5106	120	296.5265	154	380.5423
87	214.9817	121	298.9975	155	383.0133
88	217.4527	122	301.4686	156	385.4844
89	219.9238	123	303.9396	157	387.9554
90	222.3948	124	306.4107	158	390.4265
91	224.8659	125	308.8817	159	392.8976
92	227.3370	126	311.3528	160	395.3686
93	229.8080	127	313.8238	161	397.8397
94	232.2791	128	316.2949	162	400.3107
95	234.7501	129	318.7659	163	402.7818
96	237.2212	130	321.2370	164	405.2528
97	239.6922	131	323.7080	165	407.7239
98	242.1633	132	326.1791	166	410.1949
99	244.6343	133	328.6502	167	412.6660
100	247.1054	134	331.1212	168	415.1370
101	249.5764	135	333.5923	169	417.6081
102	252.0475	136	336.0633	170	420.0791
103	254.5185	137	338.5344	171	422.5502
104	256.9896	138	341.0054	172	425.0213
105	259.4606	139	343.4765	173	427.4923
106	261.9317	140	345.9475	174	429.9634
107	264.4028	141	348.4186	175	432.4344
108	266.8738	142	350.8896	176	434.9055
109	269.3449	143	353.3607	177	437.3765
110	271.8159	144	355.8317	178	439.8476
111	274.2870	145	358.3028	179	442.3186
112	276.7580	146	360.7739	180	444.7897
113	279.2291	147	363.2449	181	447.2607
114	281.7001	148	365.7160	182	449.7318
115	284.1712	149	368.1870	183	452.2028
116	286.6422	150	370.6581	184	454.6739
117	289.1133	151	373.1291	185	457.1450
118	291.5843	152	375.6002	186	459.6160
119	294.0554	153	378.0712	187	462.0871

Table 38 *(continued)*

ha.	acres	ha.	acres	ha.	acres
188	464.5581	270	667.1845	460	1136.6847
189	467.0292	280	691.8951	470	1161.3953
190	469.5002	290	716.6056	480	1186.1058
191	471.9713	300	341.3161	490	1210.8164
192	474.4423	310	766.0267	500	1235.5
193	476.9134	320	790.7372	600	1482.6
194	479.3844	330	815.4478	700	1729.7
195	481.8555	340	840.1583	800	1976.8
196	484.3265	350	864.8688	900	2223.9
197	486.7976	360	889.5794	1000	2471.1
198	489.2687	370	914.2899	2000	4942.1
199	491.7397	380	939.0004	3000	7413.2
200	494.2108	390	963.7110	4000	9884.2
210	518.9213	400	988.4215	5000	12355.3
220	543.6318	410	1013.1321	6000	14826.3
230	568.3424	420	1037.8426	7000	17297.4
240	593.0529	430	1062.5531	8000	19768.4
250	617.7635	440	1087.2637	9000	22239.5
260	642.4740	450	1111.9742	10000	24710.5

WEIGHT

Table 39 Milligrams to Grains

1 milligram = 0.154 grain

mg.	grains	mg.	grains	mg.	grains
1	.0154	33	.5093	65	1.0031
2	.0309	34	.5247	66	1.0185
3	.0463	35	.5401	67	1.0340
4	.0617	36	.5556	68	1.0494
5	.0772	37	.5710	69	1.0648
6	.0926	38	.5864	70	1.0803
7	.1080	39	.6019	71	1.0957
8	.1235	40	.6173	72	1.1111
9	.1389	41	.6327	73	1.1266
10	.1543	42	.6482	74	1.1420
11	.1698	43	.6636	75	1.1574
12	.1852	44	.6790	76	1.1729
13	.2006	45	.6945	77	1.1883
14	.2161	46	.7099	78	1.2037
15	.2315	47	.7253	79	1.2192
16	.2469	48	.7408	80	1.2346
17	.2624	49	.7562	81	1.2500
18	.2778	50	.7716	82	1.2655
19	.2932	51	.7871	83	1.2809
20	.3086	52	.8025	84	1.2963
21	.3241	53	.8179	85	1.3118
22	.3395	54	.8333	86	1.3272
23	.3549	55	.8488	87	1.3426
24	.3704	56	.8642	88	1.3580
25	.3858	57	.8796	89	1.3735
26	.4012	58	.8951	90	1.3889
27	.4167	59	.9105	91	1.4043
28	.4321	60	.9259	92	1.4198
29	.4475	61	.9414	93	1.4352
30	.4630	62	.9568	94	1.4506
31	.4784	63	.9722	95	1.4661
32	.4938	64	.9877	96	1.4815

Table 39 *(continued)*

mg.	grains	mg.	grains	mg.	grains
97	1.4969	131	2.0216	165	2.5463
98	1.5124	132	2.0371	166	2.5618
99	1.5278	133	2.0525	167	2.5772
100	1.5432	134	2.0679	168	2.5926
101	1.5587	135	2.0834	169	2.6081
102	1.5741	136	2.0988	170	2.6235
103	1.5895	137	2.1142	171	2.6389
104	1.6050	138	2.1297	172	2.6544
105	1.6204	139	2.1451	173	2.6698
106	1.6358	140	2.1605	174	2.6852
107	1.6513	141	2.1760	175	2.7007
108	1.6667	142	2.1914	176	2.7161
109	1.6821	143	2.2068	177	2.7315
110	1.6976	144	2.2223	178	2.7470
111	1.7130	145	2.2377	179	2.7624
112	1.7284	146	2.2531	180	2.7778
113	1.7439	147	2.2686	181	2.7933
114	1.7593	148	2.2840	182	2.8087
115	1.7747	149	2.2994	183	2.8241
116	1.7902	150	2.3149	184	2.8396
117	1.8056	151	2.3303	185	2.8550
118	1.8210	152	2.3457	186	2.8704
119	1.8365	153	2.3612	187	2.8859
120	1.8519	154	2.3766	188	2.9013
121	1.8673	155	2.3920	189	2.9167
122	1.8827	156	2.4074	190	2.9321
123	1.8982	157	2.4229	191	2.9476
124	1.9136	158	2.4383	192	2.9630
125	1.9290	159	2.4537	193	2.9784
126	1.9445	160	2.4692	194	2.9939
127	1.9599	161	2.4846	195	3.0093
128	1.9753	162	2.5000	196	3.0247
129	1.9908	163	2.5155	197	3.0402
130	2.0062	164	2.5309	198	3.0556

Table 39 *(continued)*

mg.	grains	mg.	grains	mg.	grains
199	3.0710	470	7.2532	740	11.420
200	3.0865	480	7.4075	750	11.574
210	3.2408	490	7.5619	760	11.729
220	3.3951	500	7.7162	770	11.883
230	3.5494	510	7.8705	780	12.037
240	3.7038	520	8.0248	790	12.192
250	3.8581	530	8.1792	800	12.346
260	4.0124	540	8.3335	810	12.500
270	4.1667	550	8.4878	820	12.655
280	4.3211	560	8.6421	830	12.809
290	4.4754	570	8.7964	840	12.963
300	4.6297	580	8.9508	850	13.118
310	4.7840	590	9.1051	860	13.272
320	4.9384	600	9.2594	870	13.426
330	5.0927	610	9.4137	880	13.580
340	5.2470	620	9.5681	890	13.735
350	5.4013	630	9.7224	900	13.889
360	5.5556	640	9.8767	910	14.043
370	5.7100	650	10.031	920	14.198
380	5.8643	660	10.185	930	14.352
390	6.0186	670	10.340	940	14.506
400	6.1729	680	10.494	950	14.661
410	6.3273	690	10.648	960	14.815
420	6.4816	700	10.803	970	14.969
430	6.6359	710	10.957	980	15.124
440	6.7902	720	11.111	990	15.278
450	6.9446	730	11.266	1000	15.432
460	7.0989				

Table 40 Grams to Avoirdupois Ounces

Note: For a comparison among weight values of avoirdupois and troy pounds and ounces, please see "Weight Conversion Tables, Customary to Customary," in Part III

1 gram = .0353 avoirdupois ounce

g.	oz.	g.	oz.	g.	oz.
1	0.03527	29	1.022945	57	2.010616
2	0.070548	30	1.058219	58	2.045890
3	0.105822	31	1.093493	59	2.081164
4	0.141096	32	1.128767	60	2.116438
5	0.176370	33	1.164041	61	2.151712
6	0.211644	34	1.199315	62	2.186986
7	0.246918	35	1.234589	63	2.222260
8	0.282192	36	1.269863	64	2.257534
9	0.317466	37	1.305137	65	2.292808
10	0.352740	38	1.340411	66	2.328081
11	0.388014	39	1.375685	67	2.363355
12	0.423388	40	1.410958	68	2.398629
13	0.458562	41	1.446232	69	2.433903
14	0.493835	42	1.481506	70	2.469177
15	0.529109	43	1.516780	71	2.504451
16	0.564383	44	1.552054	72	2.539725
17	0.599657	45	1.587328	73	2.574999
18	0.634931	46	1.622602	74	2.610273
19	0.670205	47	1.657876	75	2.645547
20	0.705479	48	1.693150	76	2.680821
21	0.740753	49	1.728424	77	2.716095
22	0.776027	50	1.763698	78	2.751369
23	0.811301	51	1.798972	79	2.786643
24	0.846575	52	1.834246	80	2.821917
25	0.881849	53	1.869520	81	2.857191
26	0.917123	54	1.904794	82	2.892465
27	0.952397	55	1.940068	83	2.927739
28	0.987671	56	1.975342	84	2.963013

Table 40 *(continued)*

g.	oz.	g.	oz.	g.	oz.
85	2.998287	119	4.197601	153	5.396916
86	3.033561	120	4.232875	154	5.432190
87	3.068835	121	4.268149	155	5.467464
88	3.104109	122	4.303423	156	5.502738
89	3.139383	123	4.338697	157	5.538012
90	3.174657	124	4.373971	158	5.573286
91	3.209931	125	4.409245	159	5.608560
92	3.245205	126	4.444519	160	5.643834
93	3.280478	127	4.479793	161	5.679108
94	3.315752	128	4.515067	162	5.714382
95	3.351026	129	4.550341	163	5.749656
96	3.386300	130	4.585615	164	5.784930
97	3.421574	131	4.620889	165	5.820204
98	3.456848	132	4.656163	166	5.855478
99	3.492122	133	4.691437	167	5.890752
100	3.527396	134	4.726711	168	5.926026
101	3.562670	135	4.761985	169	5.961300
102	3.597944	136	4.797259	170	5.996574
103	3.633218	137	4.832533	171	6.031847
104	3.668492	138	4.867807	172	6.067121
105	3.703766	139	4.903081	173	6.102395
106	3.739040	140	4.938355	174	6.137669
107	3.774314	141	4.973629	175	6.172943
108	3.809588	142	5.008903	176	6.208217
109	3.844862	143	5.044177	177	6.243491
110	3.880136	144	5.079451	178	6.278765
111	3.915410	145	5.114725	179	6.314039
112	3.950684	146	5.149998	180	6.349313
113	3.985958	147	5.185272	181	6.384587
114	4.021232	148	5.220546	182	6.419861
115	4.056506	149	5.255820	183	6.455135
116	4.091780	150	5.291094	184	6.490409
117	4.127054	151	5.326368	185	6.525683
118	4.162328	152	5.361642	186	6.560957

Table 40 *(continued)*

g.	oz.	g.	oz.	g.	oz.
187	6.596231	270	9.523970	460	16.226022
188	6.631505	280	9.876709	470	16.578762
189	6.666779	290	10.229449	480	16.931502
190	6.702053	300	10.582189	490	17.284241
191	6.737327	310	10.934928	500	17.637
192	6.772601	320	11.287668	550	19.401
193	6.807875	330	11.640407	560	19.753
194	6.843149	340	11.993147	570	20.106
195	6.878423	350	12.345887	580	20.459
196	6.913697	360	12.698626	590	20.812
197	6.948970	370	13.051366	600	21.164
198	6.984244	380	13.404106	650	22.928
199	7.019518	390	13.756845	700	24.692
200	7.054792	400	14.109585	750	26.455
210	7.407532	410	14.462324	800	28.219
220	7.760272	420	14.815064	850	29.983
230	8.113011	430	15.167804	900	31.747
240	8.465751	440	15.520542	950	33.510
250	8.818491	450	15.873283	1000	35.274
260	9.171230				

Table 41 Grams to Troy Ounces

Note: For a comparison among weight values of avoirdupois and troy pounds and ounces, please see "Weight Conversion Tables, Customary to Customary, in Part III.

1 gram = .032 troy ounce

g.	oz.	g.	oz.	g.	oz.
1	0.032151	29	0.932372	57	1.832593
2	0.064301	30	0.964522	58	1.864743
3	0.096452	31	0.996673	59	1.896894
4	0.128603	32	1.028824	60	1.929045
5	0.160754	33	1.060975	61	1.961196
6	0.192904	34	1.093125	62	1.993346
7	0.225055	35	1.125276	63	2.025497
8	0.257206	36	1.157427	64	2.057648
9	0.289357	37	1.189578	65	2.089799
10	0.321507	38	1.221728	66	2.121949
11	0.353658	39	1.253879	67	2.154100
12	0.385809	40	1.286030	68	2.186251
13	0.417960	41	1.318181	69	2.218402
14	0.450110	42	1.350331	70	2.250552
15	0.482261	43	1.382482	71	2.282703
16	0.514412	44	1.414633	72	2.314854
17	0.546563	45	1.446784	73	2.347005
18	0.578713	46	1.478934	74	2.379155
19	0.610864	47	1.511085	75	2.411306
20	0.643015	48	1.543236	76	2.443457
21	0.675166	49	1.575387	77	2.475607
22	0.707316	50	1.607537	78	2.507758
23	0.739467	51	1.639688	79	2.539909
24	0.771618	52	1.671839	80	2.572060
25	0.803769	53	1.703990	81	2.604210
26	0.835919	54	1.736140	82	2.636361
27	0.868070	55	1.768291	83	2.668512
28	0.900221	56	1.800442	84	2.700663

Table 41 *(continued)*

g.	oz.	g.	oz.	g.	oz.
85	2.732813	94	3.022170	300	9.6452
86	2.764964	95	3.054321	400	12.8603
87	2.797115	96	3.086472	500	16.0754
88	2.829266	97	3.118622	600	19.2904
89	2.861416	98	3.150773	700	22.5055
90	2.893567	99	3.182924	800	25.7206
91	2.925718	100	3.215075	900	28.9357
92	2.957869	200	6.430149	1000	32.1507
93	2.990019				

Table 42 Kilograms to Avoirdupois Pounds, Ounces

1 kilogram = 2 pounds, 3.27 ounces

kg.	lb.	oz.	kg.	lb.	oz.
½	1	1.64	26	57	5.12
1	2	3.27	27	59	8.40
2	4	6.55	28	61	11.67
3	6	9.82	29	63	14.94
4	8	14.73	30	66	2.22
5	11	.37	31	68	5.49
6	13	3.64	32	70	8.77
7	15	6.92	33	72	12.04
8	17	10.19	34	74	15.31
9	19	13.47	35	77	2.59
10	22	.74	36	79	5.86
11	24	4.01	37	81	9.14
12	26	7.29	38	83	12.41
13	28	10.56	39	85	15.68
14	30	13.84	40	88	2.96
15	33	1.11	41	90	6.23
16	35	4.38	42	92	9.51
17	37	7.66	43	94	12.78
18	39	10.93	44	97	.05
19	41	14.21	45	99	3.33
20	44	1.48	46	101	6.60
21	46	4.75	47	103	9.88
22	48	8.03	48	105	13.15
23	50	11.30	49	108	.42
24	52	14.58	50	110	3.70
25	55	1.85			

Table 43 Kilograms to Avoirdupois Pounds

1 kilogram = 2.2046 avoirdupois pounds

kg.	lb.	kg.	lb.	kg.	lb.
1	2.2046	33	72.7525	65	143.3005
2	4.4092	34	74.9572	66	145.5051
3	6.6139	35	77.1618	67	147.7097
4	8.8185	36	79.3664	68	149.9143
5	11.0231	37	81.5710	69	152.1190
6	13.2277	38	83.7757	70	154.3236
7	15.4324	39	85.9803	71	156.5282
8	17.6370	40	88.1849	72	158.7328
9	19.8416	41	90.3895	73	160.9375
10	22.0462	42	92.5942	74	163.1421
11	24.2508	43	94.7988	75	165.3467
12	26.4555	44	97.0034	76	167.5513
13	28.6601	45	99.2080	77	169.7559
14	30.8647	46	101.4126	78	171.9606
15	33.0693	47	103.6173	79	174.1652
16	35.2740	48	105.8219	80	176.3698
17	37.4786	49	108.0265	81	178.5744
18	39.6832	50	110.2311	82	180.7791
19	41.8878	51	112.4358	83	182.9837
20	44.0925	52	114.6404	84	185.1883
21	46.2971	53	116.8450	85	187.3929
22	48.5017	54	119.0496	86	189.5975
23	50.7063	55	121.2542	87	191.8022
24	52.9109	56	123.4589	88	194.0068
25	55.1156	57	125.6635	89	196.2114
26	57.3202	58	127.8681	90	198.4160
27	59.5248	59	130.0727	91	200.6207
28	61.7294	60	132.2774	92	202.8253
29	63.9341	61	134.4820	93	205.0299
30	66.1387	62	136.6866	94	207.2345
31	68.3433	63	138.8912	95	209.4391
32	70.5479	64	141.0958	96	211.6438

Table 43 *(continued)*

kg.	lb.	kg.	lb.	kg.	lb.
97	213.8484	131	288.8056	165	363.7627
98	216.0530	132	291.0102	166	365.9674
99	218.2576	133	293.2148	167	368.1720
100	220.4623	134	295.4194	168	370.3766
101	222.6669	135	297.6241	169	372.5812
102	224.8715	136	299.8287	170	374.7858
103	227.0761	137	302.0333	171	376.9905
104	229.2808	138	304.2379	172	379.1951
105	231.4854	139	306.4425	173	381.3997
106	233.6900	140	308.6472	174	383.6043
107	235.8946	141	310.8518	175	385.8090
108	238.0992	142	313.0564	176	388.0136
109	240.3039	143	315.2610	177	390.2182
110	242.5084	144	317.4657	178	392.4228
111	244.7131	145	319.6703	179	394.6274
112	246.9177	146	321.8749	180	396.8321
113	249.1224	147	324.0795	181	399.0367
114	251.3270	148	326.2841	182	401.2413
115	253.5316	149	328.4888	183	403.4459
116	255.7362	150	330.6934	184	405.6506
117	257.9408	151	332.8980	185	407.8552
118	260.1455	152	335.1026	186	410.0598
119	262.3501	153	337.3073	187	412.2644
120	264.5547	154	339.5119	188	414.4691
121	266.7593	155	341.7165	189	416.6737
122	268.9640	156	343.9211	190	418.8783
123	271.1686	157	346.1258	191	421.0829
124	273.3732	158	348.3304	192	423.2875
125	275.5778	159	350.5350	193	425.4922
126	277.7825	160	352.7396	194	427.6968
127	279.9871	161	354.9442	195	429.9014
128	282.1917	162	357.1489	196	432.1060
129	284.3963	163	359.3535	197	434.3107
130	286.6009	164	361.5581	198	436.5153

Table 43 *(continued)*

kg.	lb.	kg.	lb.	kg.	lb.
199	438.7199	365	804.6873	570	1256.635
200	440.9245	370	815.7104	580	1278.681
205	451.9476	375	826.7335	590	1300.727
210	462.9708	380	837.7566	600	1322.774
215	473.9939	385	848.7797	610	1244.820
220	485.0170	390	859.8028	620	1366.866
225	496.0401	395	870.8259	630	1388.912
230	507.0632	400	881.8490	640	1410.958
235	518.0863	405	892.8722	650	1433.005
240	529.1094	410	903.8953	660	1455.051
245	540.1325	415	914.9184	670	1477.097
250	551.1557	420	925.9415	680	1499.143
255	562.1788	425	936.9646	690	1521.190
260	573.2019	430	947.9877	700	1543.236
265	584.2250	435	959.0108	710	1565.282
270	595.2481	440	970.0340	720	1587.328
275	606.2712	445	981.0571	730	1609.375
280	617.2943	450	992.0802	740	1631.421
285	628.3174	455	1003.103	750	1653.467
290	639.3406	460	1014.126	760	1675.513
295	650.3637	465	1025.150	770	1697.559
300	661.3868	470	1036.173	780	1719.606
305	672.4099	475	1047.196	790	1741.652
310	683.4330	480	1058.219	800	1763.698
315	694.4561	485	1069.242	810	1785.744
320	705.4792	490	1080.265	820	1807.791
325	716.5024	495	1091.288	830	1829.837
330	727.5255	500	1102.311	840	1851.883
335	738.5486	510	1124.358	850	1873.929
340	749.5717	520	1146.404	860	1895.975
345	760.5948	530	1168.450	870	1918.022
350	771.6179	540	1190.496	880	1940.068
355	782.6410	550	1212.542	890	1962.114
360	793.6641	560	1234.589	900	1984.160

Table 43 *(continued)*

kg.	lb.	kg.	lb.	kg.	lb.
910	2006.207	950	2094.391	980	2160.530
920	2028.253	960	2116.438	990	2182.576
930	2050.299	970	2138.484	1000	2204.623
940	2072.345				

Table 44 Metric Tons to Short Tons (2000 Pounds)

1 metric ton = 1.1023 short tons

t.	tons	t.	tons	t.	tons
1/2	.5512	31	34.1717	62	68.3433
1	1.1023	32	35.2740	63	69.4456
2	2.2046	33	36.3763	64	70.5479
3	3.3069	34	37.4786	65	71.6502
4	4.4092	35	38.5809	66	72.7525
5	5.5116	36	39.6832	67	73.8549
6	6.6139	37	40.7855	68	74.9572
7	7.7162	38	41.8878	69	76.0595
8	8.8185	39	42.9901	70	77.1618
9	9.9208	40	44.0925	71	78.2641
10	11.0231	41	45.1948	72	79.3664
11	12.1254	42	46.2971	73	80.4687
12	13.2277	43	47.3994	74	81.5710
13	14.3300	44	48.5017	75	82.6733
14	15.4324	45	49.6040	76	83.7757
15	16.5347	46	50.7063	77	84.8780
16	17.6370	47	51.8086	78	85.9803
17	18.7393	48	52.9109	79	87.0826
18	19.8416	49	54.0133	80	88.1849
19	20.3928	50	55.1156	81	89.2872
20	22.0462	51	56.2179	82	90.3894
21	23.1485	52	57.3202	83	91.4918
22	24.2508	53	58.4225	84	92.5941
23	25.3532	54	59.5248	85	93.6965
24	26.4555	55	60.6271	86	94.7988
25	27.5578	56	61.7294	87	95.9011
26	28.6601	57	62.8317	88	97.0034
27	29.7624	58	63.9341	89	98.1057
28	30.8647	59	65.0364	90	99.2080
29	31.9670	60	66.1387	91	100.310
30	33.0693	61	67.2410	92	101.413

Table 44 *(continued)*

t.	tons	t.	tons	t.	tons
93	102.515	127	139.994	161	177.472
94	103.617	128	141.096	162	178.574
95	104.720	129	142.198	163	179.677
96	105.822	130	143.300	164	180.779
97	106.924	131	144.403	165	181.881
98	108.027	132	145.505	166	182.984
99	109.129	133	146.607	167	184.086
100	110.231	134	147.710	168	185.188
101	111.333	135	148.812	169	186.291
102	112.436	136	149.914	170	187.393
103	113.538	137	151.017	171	188.495
104	114.640	138	152.119	172	189.598
105	115.743	139	153.221	173	190.700
106	116.845	140	154.324	174	191.802
107	117.947	141	155.426	175	192.904
108	119.050	142	156.528	176	194.007
109	120.152	143	157.631	177	195.109
110	121.254	144	158.733	178	196.211
111	122.357	145	159.835	179	197.314
112	123.459	146	160.937	180	198.416
113	124.561	147	162.040	181	199.518
114	125.663	148	163.142	182	200.621
115	126.766	149	164.244	183	201.723
116	127.868	150	165.347	184	202.825
117	128.970	151	166.449	185	203.928
118	130.073	152	167.551	186	205.030
119	131.175	153	168.654	187	206.132
120	132.277	154	169.756	188	207.235
121	133.380	155	170.858	189	208.337
122	134.482	156	171.961	190	209.439
123	135.584	157	173.063	191	210.541
124	136.687	158	174.165	192	211.644
125	137.789	159	175.267	193	212.746
126	138.891	160	176.370	194	213.848

Table 44 *(continued)*

t.	tons	t.	tons	t.	tons
195	214.951	470	518.086	790	870.826
196	216.053	480	529.109	800	881.849
197	217.155	490	540.133	810	892.872
198	218.258	500	551.156	820	903.895
199	219.360	510	562.179	830	914.918
200	220.462	520	573.202	840	925.941
210	231.485	530	584.225	850	936.965
220	242.508	540	595.248	860	947.988
230	253.532	550	606.271	870	959.011
240	264.555	560	617.294	880	970.034
250	275.578	570	628.317	890	981.057
260	286.601	580	639.341	900	992.080
270	297.624	590	650.364	910	1003.10
280	308.647	600	661.387	920	1014.13
290	319.670	610	672.410	930	1025.15
300	330.693	620	683.433	940	1036.17
310	341.717	630	694.456	950	1047.20
320	352.740	640	705.479	960	1058.22
330	363.763	650	716.502	970	1069.24
340	374.786	660	727.525	980	1080.27
350	385.809	670	738.549	990	1091.29
360	396.832	680	749.572	1000	1102.31
370	407.855	690	760.595	2000	2204.62
380	418.878	700	771.618	3000	3306.93
390	429.901	710	782.641	4000	4409.25
400	440.925	720	793.664	5000	5511.56
410	451.948	730	804.687	6000	6613.87
420	462.971	740	815.710	7000	7716.18
430	473.994	750	826.733	8000	8818.49
440	485.017	760	837.757	9000	9920.80
450	496.040	770	848.780	10000	11023.11
460	507.063	780	859.803		

CAPACITY

Table 45 Milliliters to Teaspoons and Tablespoons

1 milliliter = .203 teaspoon and .068 tablespoon

ml.	tspn.	tbspn.	ml.	tspn.	tbspn.
1	.20	.07	32	6.49	2.16
2	.41	.14	33	6.70	2.23
3	.61	.20	34	6.90	2.30
4	.81	.27	35	7.10	2.37
5	1.01	.34	36	7.30	2.43
6	1.22	.41	37	7.51	2.50
7	1.42	.47	38	7.71	2.57
8	1.62	.54	39	7.91	2.64
9	1.83	.61	40	8.12	2.71
10	2.03	.68	41	8.32	2.77
11	2.23	.74	42	8.52	2.84
12	2.43	.81	43	8.72	2.91
13	2.64	.88	44	8.93	2.98
14	2.84	.95	45	9.13	3.04
15	3.04	1.01	46	9.33	3.11
16	3.25	1.08	47	9.54	3.18
17	3.45	1.15	48	9.74	3.25
18	3.65	1.22	49	9.94	3.31
19	3.85	1.28	50	10.14	3.38
20	4.06	1.35	51	10.35	3.45
21	4.26	1.42	52	10.55	3.52
22	4.46	1.49	53	10.75	3.58
23	4.67	1.56	54	10.96	3.65
24	4.87	1.62	55	11.16	3.72
25	5.07	1.69	56	11.36	3.79
26	5.27	1.76	57	11.56	3.85
27	5.48	1.83	58	11.77	3.92
28	5.68	1.89	59	11.97	3.99
29	5.88	1.96	60	12.17	4.06
30	6.09	2.03	61	12.38	4.13
31	6.29	2.10	62	12.58	4.19

Table 45 *(continued)*

ml.	tspn.	tbspn.	ml.	tspn.	tbspn.
63	12.78	4.26	82	16.64	5.55
64	12.98	4.33	83	16.84	5.61
65	13.19	4.40	84	17.04	5.68
66	13.39	4.46	85	17.25	5.75
67	13.59	4.53	86	17.45	5.82
68	13.80	4.60	87	17.65	5.88
69	14.00	4.67	88	17.85	5.95
70	14.20	4.73	89	18.06	6.02
71	14.40	4.80	90	18.26	6.09
72	14.61	4.87	91	18.46	6.15
73	14.81	4.94	92	18.67	6.22
74	15.01	5.00	93	18.87	6.29
75	15.22	5.07	94	19.07	6.36
76	15.42	5.14	95	19.27	6.42
77	15.62	5.21	96	19.48	6.49
78	15.82	5.27	97	19.68	6.56
79	16.03	5.34	98	19.88	6.63
80	16.23	5.41	99	20.09	6.70
81	16.43	5.48	100	20.29	6.76

Table 46 Milliliters to Liquid Ounces

1 milliliter = .0336 liquid ounce

ml.	oz.	ml.	oz.	ml.	oz.
1	0.03381	33	1.11586	65	2.19791
2	0.06763	34	1.14968	66	2.23173
3	0.10144	35	1.18349	67	2.26554
4	0.13526	36	1.21730	68	2.29935
5	0.16907	37	1.25112	69	2.33317
6	0.20288	38	1.28493	70	2.36698
7	0.23670	39	1.31875	71	2.40080
8	0.27051	40	1.35256	72	2.43461
9	0.30433	41	1.38637	73	2.46842
10	0.33814	42	1.42019	74	2.50224
11	0.37195	43	1.45400	75	2.53605
12	0.40577	44	1.48782	76	2.56987
13	0.43958	45	1.52163	77	2.60368
14	0.47340	46	1.55544	78	2.63749
15	0.50721	47	1.58926	79	2.67131
16	0.54102	48	1.62307	80	2.70512
17	0.57484	49	1.65689	81	2.73894
18	0.60865	50	1.69070	82	2.77275
19	0.64247	51	1.72452	83	2.80656
20	0.67628	52	1.75833	84	2.84038
21	0.71009	53	1.79214	85	2.87419
22	0.74391	54	1.82596	86	2.90801
23	0.77772	55	1.85977	87	2.94182
24	0.81154	56	1.89359	88	2.97563
25	0.84535	57	1.92740	89	3.00945
26	0.87916	58	1.96121	90	3.04326
27	0.91298	59	1.99503	91	3.07708
28	0.94679	60	2.02884	92	3.11089
29	0.98061	61	2.06266	93	3.14470
30	1.01442	62	2.09647	94	3.17852
31	1.04823	63	2.13028	95	3.21233
32	1.08205	64	2.16410	96	3.24615

Table 46 *(continued)*

ml.	oz.	ml.	oz.	ml.	oz.
97	3.27996	410	13.864	750	25.361
98	3.31377	420	14.202	760	25.699
99	3.34759	430	14.540	770	26.037
100	3.381	440	14.878	780	26.375
110	3.720	450	15.216	790	26.713
120	4.058	460	15.554	800	27.051
130	4.396	470	15.893	810	27.389
140	4.734	480	16.231	820	27.727
150	5.072	490	16.569	830	28.066
160	5.410	500	16.907	840	28.404
170	5.748	510	17.245	850	28.742
180	6.087	520	17.583	860	29.080
190	6.425	530	17.921	870	29.418
200	6.763	540	18.260	880	29.756
210	7.101	550	18.598	890	30.094
220	7.439	560	18.936	900	30.433
230	7.777	570	19.274	910	30.771
240	8.115	580	19.612	920	31.109
250	8.454	590	19.950	930	31.447
260	8.792	600	20.288	940	31.785
270	9.130	610	20.627	950	32.123
280	9.468	620	20.965	960	32.461
290	9.806	630	21.303	970	32.800
300	10.144	640	21.641	980	33.138
310	10.482	650	21.979	990	33.476
320	10.820	660	22.317	1000	33.814
330	11.159	670	22.655	1100	37.195
340	11.497	680	22.994	1200	40.577
350	11.835	690	23.332	1300	43.958
360	12.173	700	23.670	1400	47.340
370	12.511	710	24.008	1500	50.721
380	12.849	720	24.346	1600	54.102
390	13.187	730	24.684	1700	57.484
400	13.526	740	25.022	1800	60.865

Table 46 *(continued)*

ml.	oz.	ml.	oz.	ml.	oz.
1900	64.247	2300	77.772	2700	91.298
2000	67.628	2400	81.154	2800	94.679
2100	71.009	2500	84.535	2900	98.061
2200	74.391	2600	87.916	3000	101.442

Table 47 Liters to Liquid Quarts

1 liter = 1.056 liquid quarts

l.	qt.	l.	qt.	l.	qt.
0.1	.1057	22	23.2471	54	57.0612
0.2	.2113	23	24.3038	55	58.1179
0.25	.2642	24	25.3605	56	59.1745
0.3	.3170	25	26.4172	57	60.2312
0.4	.4227	26	27.4739	58	61.2879
0.5	.5283	27	28.5306	59	62.3446
0.6	.6340	28	29.5873	60	63.4013
0.7	.7397	29	30.6440	61	64.4580
0.75	.7925	30	31.7006	62	65.5147
0.8	.8454	31	32.7573	63	66.5714
0.9	.9510	32	33.8140	64	67.6280
1	1.0567	33	34.8707	65	68.6847
2	2.1134	34	35.9274	66	69.7414
3	3.1701	35	36.9841	67	70.7981
4	4.2268	36	38.0408	68	71.8548
5	5.2834	37	39.0975	69	72.9115
6	6.3401	38	40.1542	70	73.9682
7	7.3968	39	41.2108	71	75.0249
8	8.4535	40	42.2675	72	76.0816
9	9.5102	41	43.3242	73	77.1382
10	10.5669	42	44.3809	74	78.1949
11	11.6236	43	45.4376	75	79.2516
12	12.6803	44	46.4943	76	80.3083
13	13.7369	45	47.5510	77	81.3650
14	14.7936	46	48.6077	78	82.4217
15	15.8503	47	49.6643	79	83.4784
16	16.9070	48	50.7210	80	84.5351
17	17.9637	49	51.7777	81	85.5917
18	19.0204	50	52.8344	82	86.6484
19	20.0771	51	53.8911	83	87.7051
20	21.1338	52	54.9478	84	88.7618
21	22.1905	53	56.0045	85	89.8185

Table 47 *(continued)*

l.	qt.	l.	qt.	l.	qt.
86	90.8752	160	169.0701	360	380.4078
87	91.9319	170	179.6370	370	390.9746
88	92.9886	180	190.2039	380	401.5415
89	94.0452	190	200.7708	390	412.1084
90	95.1019	200	211.3376	400	422.6753
91	96.1586	210	221.9045	410	433.2422
92	97.2153	220	232.4714	420	443.8090
93	98.2720	230	243.0383	430	454.3759
94	99.3287	240	253.6052	440	464.9428
95	100.3854	250	264.1721	450	475.5097
96	101.4421	260	274.7389	460	486.0766
97	102.4988	270	285.3058	470	496.6435
98	103.5554	280	295.8727	480	507.2103
99	104.6121	290	306.4396	490	517.7772
100	105.6688	300	317.0065	500	528.34
110	116.2357	310	327.5733	600	634.01
120	126.8026	320	338.1402	700	739.68
130	137.3695	330	348.7071	800	845.35
140	147.9363	340	359.2740	900	951.02
150	158.5032	350	369.8409	1000	1056.69

Table 48 Liters to Gallons

1 liter = .2642 gallon

l.	gal.	l.	gal.	l.	gal.
1	.264	33	8.718	65	17.171
2	.528	34	8.982	66	17.435
3	.793	35	9.246	67	17.700
4	1.057	36	9.510	68	17.964
5	1.321	37	9.774	69	18.228
6	1.585	38	10.039	70	18.492
7	1.849	39	10.303	71	18.756
8	2.113	40	10.567	72	19.020
9	2.378	41	10.831	73	19.285
10	2.642	42	11.095	74	19.549
11	2.906	43	11.359	75	19.813
12	3.170	44	11.624	76	20.077
13	3.434	45	11.888	77	20.341
14	3.698	46	12.152	78	20.605
15	3.963	47	12.416	79	20.870
16	4.227	48	12.680	80	21.134
17	4.491	49	12.944	81	21.398
18	4.755	50	13.209	82	21.662
19	5.019	51	13.473	83	21.926
20	5.283	52	13.737	84	22.190
21	5.548	53	14.001	85	22.455
22	5.812	54	14.265	86	22.719
23	6.076	55	14.529	87	22.983
24	6.340	56	14.794	88	23.247
25	6.604	57	15.058	89	23.511
26	6.868	58	15.322	90	23.775
27	7.133	59	15.586	91	24.040
28	7.397	60	15.850	92	24.304
29	7.661	61	16.114	93	24.568
30	7.925	62	16.379	94	24.832
31	8.189	63	16.643	95	25.096
32	8.454	64	16.907	96	25.361

Table 48 *(continued)*

l.	gal.	l.	gal.	l.	gal.
97	25.625	131	34.607	165	43.588
98	25.889	132	34.871	166	43.853
99	26.153	133	35.135	167	44.117
100	26.417	134	35.399	168	44.381
101	26.681	135	35.663	169	44.645
102	26.946	136	35.927	170	44.909
103	27.210	137	36.192	171	45.173
104	27.474	138	36.456	172	45.438
105	27.738	139	36.720	173	45.702
106	28.002	140	36.984	174	45.966
107	28.266	141	37.248	175	46.230
108	28.531	142	37.512	176	46.494
109	28.795	143	37.777	177	46.758
110	29.059	144	38.041	178	47.023
111	29.323	145	38.305	179	47.287
112	29.587	146	38.569	180	47.551
113	29.851	147	38.833	181	47.815
114	30.116	148	39.097	182	48.079
115	30.380	149	39.362	183	48.343
116	30.644	150	39.626	184	46.608
117	30.908	151	39.890	185	48.872
118	31.172	152	40.154	186	49.136
119	31.436	153	40.418	187	49.400
120	31.701	154	40.682	188	49.664
121	31.965	155	40.947	189	49.929
122	32.229	156	41.211	190	50.193
123	32.493	157	41.475	191	50.457
124	32.757	158	41.739	192	50.721
125	33.022	159	42.003	193	50.985
126	33.286	160	42.268	194	51.249
127	33.550	161	42.532	195	51.514
128	33.814	162	42.796	196	51.778
129	34.078	163	43.060	197	52.042
130	34.342	164	43.324	198	52.306

Table 48 *(continued)*

l.	gal.	l.	gal.	l.	gal.
199	52.570	480	126.803	770	203.412
200	52.834	490	129.444	780	206.054
210	55.476	500	132.086	790	208.696
220	58.118	510	134.728	800	211.338
230	60.760	520	137.369	810	213.979
240	63.401	530	140.011	820	216.621
250	66.043	540	142.653	830	219.263
260	68.685	550	145.295	840	221.905
270	71.326	560	147.936	850	224.546
280	73.968	570	150.578	860	227.188
290	76.610	580	153.220	870	229.830
300	79.252	590	155.862	880	232.471
310	81.893	600	158.503	890	235.113
320	84.535	610	161.145	900	237.755
330	87.177	620	163.787	910	240.397
340	89.818	630	166.428	920	243.038
350	92.460	640	169.070	930	245.680
360	95.102	650	171.712	940	248.322
370	97.744	660	174.354	950	250.963
380	100.385	670	176.995	960	253.605
390	103.027	680	179.637	970	256.247
400	105.669	690	182.279	980	258.889
410	108.311	700	184.920	990	261.530
420	110.952	710	187.562	1000	264.172
430	113.594	720	190.204	2000	528.344
440	116.236	730	192.846	3000	792.516
450	118.877	740	195.487	4000	1056.688
460	121.519	750	198.129	5000	1320.860
470	124.161	760	200.771		

Table 49 Liters to Gallons, Liquid Quarts, Liquid Pints

1 liter = 0 gallons, 1 liquid quart, .1134 liquid pint

l.	gal.	qt.	pt.	l.	gal.	qt.	pt.
0.1			.21	21	5	2	.38
0.2			.42	22	5	3	.49
0.25			.53	23	6		.61
0.3			.63	24	6	1	.72
0.4			.85	25	6	2	.83
0.5			1.06	26	6	3	.95
0.6			1.27	27	7		1.06
0.7			1.48	28	7	1	1.17
0.75			1.59	29	7	2	1.29
0.8			1.69	30	7	3	1.40
0.9			1.90	31	8		1.51
1		1	.12	32	8	1	1.63
2		2	.23	33	8	2	1.74
3		3	.34	34	8	3	1.85
4	1		.45	35	9		1.97
5	1	1	.57	36	9	2	.08
6	1	2	.68	37	9	3	.19
7	1	3	.79	38	10		.31
8	2		.91	39	10	1	.42
9	2	1	1.02	40	10	2	.54
10	2	2	1.13	41	10	3	.65
11	2	3	1.25	42	11		.76
12	3		1.36	43	11	1	.88
13	3	1	1.47	44	11	2	.99
14	3	2	1.59	45	11	3	1.10
15	3	3	1.70	46	12		1.22
16	4		1.81	47	12	1	1.33
17	4	1	1.93	48	12	2	1.44
18	4	3	.04	49	12	3	1.56
19	5		.15	50	13		1.67
20	5	1	.27	51	13	1	1.78

Table 49 *(continued)*

l.	gal.	qt.	pt.	l.	gal.	qt.	pt.
52	13	2	1.90	77	20	1	.73
53	14		.01	78	20	2	.84
54	14	1	.12	79	20	3	.96
55	14	2	.24	80	21		1.07
56	14	3	.35	81	21	1	1.18
57	15		.46	82	21	2	1.30
58	15	1	.58	83	21	3	1.41
59	15	2	.69	84	22		1.52
60	15	3	.80	85	22	1	1.64
61	16		.92	86	22	2	1.75
62	16	1	1.03	87	22	3	1.86
63	16	2	1.14	88	23		1.98
64	16	3	1.26	89	23	2	.09
65	17		1.37	90	23	3	.20
66	17	1	1.48	91	24		.32
67	17	2	1.60	92	24	1	.43
68	17	3	1.71	93	24	2	.54
69	18		1.82	94	24	3	.66
70	18	1	1.94	95	25		.77
71	18	3	.05	96	25	1	.88
72	19		.16	97	25	2	1.00
73	19	1	.28	98	25	3	1.11
74	19	2	.39	99	26		1.22
75	19	3	.50	100	26	1	1.34
76	20		.62				

Table 50 Liters to Bushels, Pecks

1 liter = 0 bushels, .114 peck

l.	bu.	pecks	l.	bu.	pecks
1		.114	33		3.75
2		.227	34		3.86
3		.341	35		3.97
4		.454	36	1	.09
5		.568	37	1	.20
6		.681	38	1	.31
7		.795	39	1	.43
8		.908	40	1	.54
9		1.02	41	1	.65
10		1.13	42	1	.77
11		1.25	43	1	.88
12		1.36	44	1	.99
13		1.48	45	1	1.11
14		1.59	46	1	1.22
15		1.70	47	1	1.33
16		1.82	48	1	1.45
17		1.93	49	1	1.56
18		2.04	50	1	1.68
19		2.16	51	1	1.79
20		2.27	52	1	1.90
21		2.38	53	1	2.02
22		2.50	54	1	2.13
23		2.61	55	1	2.24
24		2.72	56	1	2.36
25		2.84	57	1	2.47
26		2.95	58	1	2.58
27		3.06	59	1	2.70
28		3.18	60	1	2.81
29		3.29	61	1	2.92
30		3.41	62	1	3.04
31		3.52	63	1	3.15
32		3.63	64	1	3.26

Table 50 *(continued)*

l.	bu.	pecks	l.	bu.	pecks
65	1	3.38	99	2	3.24
66	1	3.49	100	2	3.35
67	1	3.61	101	2	3.46
68	1	3.72	102	2	3.58
69	1	3.83	103	2	3.69
70	1	3.95	104	2	3.81
71	2	.06	105	2	3.92
72	2	.17	106	3	.03
73	2	.29	107	3	.15
74	2	.40	108	3	.26
75	2	.51	109	3	.37
76	2	.63	110	3	.49
77	2	.74	111	3	.60
78	2	.85	112	3	.71
79	2	.97	113	3	.83
80	2	1.08	114	3	.94
81	2	1.19	115	3	1.05
82	2	1.31	116	3	1.17
83	2	1.42	117	3	1.28
84	2	1.53	118	3	1.39
85	2	1.65	119	3	1.51
86	2	1.76	120	3	1.62
87	2	1.88	121	3	1.73
88	2	1.99	122	3	1.85
89	2	2.10	123	3	1.96
90	2	2.22	124	3	2.08
91	2	2.33	125	3	2.19
92	2	2.44	126	3	2.30
93	2	2.56	127	3	2.42
94	2	2.67	128	3	2.53
95	2	2.78	129	3	2.64
96	2	2.90	130	3	2.76
97	2	3.01	131	3	2.87
98	2	3.12	132	3	2.98

Table 50 *(continued)*

l.	bu.	pecks	l.	bu.	pecks
133	3	3.10	167	4	2.96
134	3	3.21	168	4	3.07
135	3	3.32	169	4	3.18
136	3	3.44	170	4	3.30
137	3	3.55	171	4	3.41
138	3	3.66	172	4	3.52
139	3	3.78	173	4	3.64
140	3	3.89	174	4	3.75
141	4	.00	175	4	3.86
142	4	.12	176	4	3.98
143	4	.23	177	5	.09
144	4	.35	178	5	.20
145	4	.46	179	5	.32
146	4	.57	180	5	.43
147	4	.69	181	5	.55
148	4	.80	182	5	.66
149	4	.91	183	5	.77
150	4	1.03	184	5	.89
151	4	1.14	185	5	1.00
152	4	1.25	186	5	1.11
153	4	1.37	187	5	1.23
154	4	1.48	188	5	1.34
155	4	1.59	189	5	1.45
156	4	1.71	190	5	1.57
157	4	1.82	191	5	1.68
158	4	1.93	192	5	1.79
159	4	2.05	193	5	1.91
160	4	2.16	194	5	2.02
161	4	2.28	195	5	2.13
162	4	2.39	196	5	2.25
163	4	2.50	197	5	2.36
164	4	2.62	198	5	2.48
165	4	2.73	199	5	2.59
166	4	2.84	200	5	2.70

Table 51 Liters to Bushels

1 liter = .0284 bushel

l.	bu.	l.	bu.	l.	bu.
200	5.67552	232	6.58360	264	7.49168
201	5.70390	233	6.61198	265	7.52006
202	5.73227	234	6.64036	266	7.54844
203	5.76065	235	6.66873	267	7.57682
204	5.78903	236	6.69711	268	7.60519
205	5.81741	237	6.72549	269	7.63357
206	5.84578	238	6.75387	270	7.66195
207	5.87416	239	6.78224	271	7.69033
208	5.90254	240	6.81062	272	7.71870
209	5.93092	241	6.83900	273	7.74708
210	5.95929	242	6.86738	274	7.77546
211	5.98767	243	6.89575	275	7.80384
212	6.01605	244	6.92413	276	7.83221
213	6.04443	245	6.95251	277	7.86059
214	6.07280	246	6.98089	278	7.88897
215	6.10118	247	7.00926	279	7.91735
216	6.12956	248	7.03764	280	7.94573
217	6.15794	249	7.06602	281	7.97410
218	6.18631	250	7.09440	282	8.00248
219	6.21469	251	7.12278	283	8.03086
220	6.24307	252	7.15115	284	8.05924
221	6.27145	253	7.17953	285	8.08761
222	6.29982	254	7.20791	286	8.11599
223	6.32820	255	7.23629	287	8.14437
224	6.35658	256	7.26466	288	8.17275
225	6.38496	257	7.29304	289	8.20112
226	6.41334	258	7.32142	290	8.22950
227	6.44171	259	7.34980	291	8.25788
228	6.47009	260	7.37817	292	8.28626
229	6.49847	261	7.40655	293	8.31463
230	6.52685	262	7.43493	294	8.34301
231	6.55522	263	7.46331	295	8.37139

Table 51 *(continued)*

l.	bu.	l.	bu.	l.	bu.
296	8.39977	455	12.912	760	21.567
297	8.42814	460	13.054	770	21.851
298	8.45652	465	13.196	780	22.135
299	8.48490	470	13.337	790	22.418
300	8.5133	475	13.479	800	22.702
305	8.6552	480	13.621	810	22.986
310	8.7971	485	13.763	820	23.270
315	8.9389	490	13.905	830	23.553
320	9.0808	495	14.047	840	23.837
325	9.2227	500	14.189	850	24.121
330	9.3646	510	14.473	860	24.405
335	9.5065	520	14.756	870	24.689
340	9.6484	530	15.040	880	24.972
345	9.7903	540	15.324	890	25.256
350	9.9322	550	15.608	900	25.540
355	10.074	560	15.891	910	25.824
360	10.216	570	16.175	920	26.107
365	10.358	580	16.459	930	26.391
370	10.500	590	16.743	940	26.675
375	10.642	600	17.027	950	26.959
380	10.783	610	17.310	960	27.242
385	10.925	620	17.594	970	27.526
390	11.067	630	17.878	980	27.810
395	11.209	640	18.162	990	28.094
400	11.351	650	18.445	1000	28.378
405	11.493	660	18.729	2000	56.76
410	11.635	670	19.013	3000	85.13
415	11.777	680	19.297	4000	113.51
420	11.919	690	19.581	5000	141.89
425	12.060	700	19.864	6000	170.27
430	12.202	710	20.148	7000	198.64
435	12.344	720	20.432	8000	227.02
440	12.486	730	20.716	9000	255.40
445	12.628	740	20.999	10000	283.78
450	12.770	750	21.283		

VOLUME

Table 52 Cubic Centimeters to Cubic Inches

1 cubic centimeter = .061 cubic inch

cm.^{3}	in.^{3}	cm.^{3}	in.^{3}	cm.^{3}	in.^{3}
1	.061	33	2.014	65	3.967
2	.122	34	2.075	66	4.028
3	.183	35	2.136	67	4.089
4	.244	36	2.197	68	4.150
5	.305	37	2.258	69	4.211
6	.366	38	2.319	70	4.272
7	.427	39	2.380	71	4.333
8	.488	40	2.441	72	4.394
9	.549	41	2.502	73	4.455
10	.610	42	2.563	74	4.516
11	.671	43	2.624	75	4.577
12	.732	44	2.685	76	4.638
13	.793	45	2.746	77	4.699
14	.854	46	2.807	78	4.760
15	.915	47	2.868	79	4.821
16	.976	48	2.929	80	4.882
17	1.037	49	2.990	81	4.943
18	1.098	50	3.051	82	5.004
19	1.159	51	3.112	83	5.065
20	1.220	52	3.173	84	5.126
21	1.281	53	3.234	85	5.187
22	1.343	54	3.295	86	5.248
23	1.404	55	3.356	87	5.309
24	1.465	56	3.417	88	5.370
25	1.526	57	3.478	89	5.431
26	1.587	58	3.539	90	5.492
27	1.648	59	3.600	91	5.553
28	1.709	60	3.661	92	5.614
29	1.770	61	3.722	93	5.675
30	1.831	62	3.783	94	5.736
31	1.892	63	3.844	95	5.797
32	1.953	64	3.906	96	5.858

Table 52 *(continued)*

cm.^{3}	in.^{3}	cm.^{3}	in.^{3}	cm.^{3}	in.^{3}
97	5.919	131	7.994	165	10.069
98	5.980	132	8.055	166	10.130
99	6.041	133	8.116	167	10.191
100	6.102	134	8.177	168	10.252
101	6.163	135	8.238	169	10.313
102	6.224	136	8.299	170	10.374
103	6.285	137	8.360	171	10.435
104	6.346	138	8.421	172	10.496
105	6.407	139	8.482	173	10.557
106	6.469	140	8.543	174	10.618
107	6.530	141	8.604	175	10.679
108	6.591	142	8.665	176	10.740
109	6.652	143	8.726	177	10.801
110	6.713	144	8.787	178	10.862
111	6.774	145	8.848	179	10.923
112	6.835	146	8.909	180	10.984
113	6.896	147	8.970	181	11.045
114	6.957	148	9.032	182	11.106
115	7.018	149	9.093	183	11.167
116	7.079	150	9.154	184	11.228
117	7.140	151	9.215	185	11.289
118	7.201	152	9.276	186	11.350
119	7.262	153	9.337	187	11.411
120	7.323	154	9.398	188	11.472
121	7.384	155	9.459	189	11.533
122	7.445	156	9.520	190	11.595
123	7.506	157	9.581	191	11.656
124	7.567	158	9.642	192	11.717
125	7.628	159	9.703	193	11.778
126	7.689	160	9.764	194	11.839
127	7.750	161	9.825	195	11.900
128	7.811	162	9.886	196	11.961
129	7.872	163	9.947	197	12.022
130	7.933	164	10.008	198	12.083

Table 52 *(continued)*

cm.^{3}	in.^{3}	cm.^{3}	in.^{3}	cm.^{3}	in.^{3}
199	12.144	500	30.512	800	48.819
200	12.205	510	31.122	810	49.429
210	12.815	520	31.732	820	50.039
220	13.425	530	32.343	830	50.650
230	14.035	540	32.953	840	51.260
240	14.646	550	33.563	850	51.870
250	15.256	560	34.173	860	52.480
260	15.866	570	34.784	870	53.091
270	16.476	580	35.394	880	53.701
280	17.087	590	36.004	890	54.311
290	17.697	600	36.614	900	54.921
300	18.307	610	37.224	910	55.532
310	18.917	620	37.835	920	56.142
320	19.528	630	38.445	930	56.752
330	20.138	640	39.055	940	57.362
340	20.748	650	39.665	950	57.973
350	21.358	660	40.276	960	58.583
360	21.969	670	40.886	970	59.193
370	22.579	680	41.496	980	59.803
380	23.189	690	42.106	990	60.414
390	23.799	700	42.717	1000	61.024
400	24.409	710	43.327	2000	122.047
410	25.020	720	43.937	3000	183.071
420	25.630	730	44.547	4000	244.095
430	26.240	740	45.158	5000	305.119
440	26.850	750	45.768	6000	366.142
450	27.461	760	46.378	7000	427.166
460	28.071	770	46.988	8000	488.190
470	28.681	780	47.599	9000	549.214
480	29.291	790	48.209	10000	610.237
490	29.902				

Table 53 Cubic Meters to Cubic Feet and Cubic Yards

1 cubic meter = 35.315 cubic feet and 1.308 cubic yards

$m.^3$	$ft.^3$	$yd.^3$	$m.^3$	$ft.^3$	$yd.^3$
1	35.31	1.31	32	1130.07	41.85
2	70.63	2.62	33	1165.38	43.16
3	105.94	3.92	34	1200.70	44.47
4	141.26	5.23	35	1236.01	45.78
5	176.57	6.54	36	1271.33	47.09
6	211.89	7.85	37	1306.64	48.39
7	247.20	9.16	38	1341.96	49.70
8	282.52	10.46	39	1377.27	51.01
9	317.83	11.77	40	1412.59	52.32
10	353.15	13.08	41	1447.90	53.63
11	388.46	14.39	42	1483.22	54.93
12	423.78	15.70	43	1518.53	56.24
13	459.09	17.00	44	1553.85	57.55
14	494.41	18.31	45	1589.16	58.86
15	529.72	19.62	46	1624.47	60.17
16	565.03	20.93	47	1659.79	61.47
17	600.35	22.24	48	1695.10	62.78
18	635.66	23.54	49	1730.42	60.09
19	670.98	24.85	50	1765.73	65.40
20	706.29	26.16	51	1801.05	66.71
21	741.61	27.47	52	1836.36	68.01
22	776.92	28.77	53	1871.68	69.32
23	812.24	30.08	54	1906.99	70.63
24	847.55	31.39	55	1942.31	71.94
25	882.87	32.70	56	1977.62	73.25
26	918.18	34.01	57	2012.94	74.55
27	953.50	35.31	58	2048.25	74.86
28	988.81	36.62	59	2083.57	77.17
29	1024.13	37.93	60	2118.88	78.48
30	1059.44	39.24	61	2154.19	79.78
31	1094.75	40.55	62	2189.51	81.09

Table 53 *(continued)*

m.^{3}	ft.^{3}	yd.^{3}	m.^{3}	ft.^{3}	yd.^{3}
63	2224.82	82.40	97	3425.52	126.87
64	2260.14	83.71	98	3460.84	128.18
65	2295.45	85.02	99	3496.15	129.49
66	2330.77	86.32	100	3531.47	130.80
67	2366.08	87.63	101	3566.78	132.10
68	2401.40	88.94	102	3602.10	133.41
69	2436.71	90.25	103	3637.41	134.72
70	2472.03	91.56	104	3672.73	136.03
71	2507.34	92.86	105	3708.04	137.33
72	2542.66	94.17	106	3743.35	138.64
73	2577.97	95.48	107	3778.67	139.95
74	2613.29	96.79	108	3813.98	141.26
75	2648.60	98.10	109	3849.30	142.57
76	2683.91	99.40	110	3884.61	143.87
77	2719.23	100.71	111	3919.93	145.18
78	2754.54	102.02	112	3955.24	146.49
79	2789.86	103.33	113	3990.56	147.80
80	2825.17	104.64	114	4025.87	149.11
81	2860.49	105.94	115	4061.19	150.41
82	2895.80	107.25	116	4096.50	151.72
83	2931.12	108.56	117	4131.82	153.03
84	2966.43	109.87	118	4167.13	154.34
85	3001.75	111.18	119	4202.45	155.65
86	3037.06	112.48	120	4237.76	156.95
87	3072.38	113.79	121	4273.07	158.26
88	3107.69	115.10	122	4308.39	159.57
89	3143.01	116.41	123	4344.70	160.88
90	3178.32	117.72	124	4379.02	162.19
91	3213.63	119.02	125	4414.33	163.49
92	3248.95	120.33	126	4449.65	164.80
93	3284.26	121.64	127	4484.96	166.11
94	3319.58	122.95	128	4520.28	167.42
95	3354.89	124.26	129	4555.59	168.73
96	3390.21	125.56	130	4590.91	170.03

Table 53 *(continued)*

m.^{3}	ft.^{3}	yd.^{3}	m.^{3}	ft.^{3}	yd.^{3}
131	4626.22	171.34	165	5826.92	215.81
132	4661.54	172.65	166	5862.23	217.12
133	4696.85	173.96	167	5897.55	218.43
134	4732.17	175.27	168	5932.86	219.74
135	4767.48	176.57	169	5968.18	221.04
136	4802.79	177.88	170	6003.49	222.35
137	4838.11	179.19	171	6038.81	223.66
138	4873.42	180.50	172	6074.12	224.97
139	4908.74	181.81	173	6109.44	226.28
140	4944.05	183.11	174	6144.75	227.58
141	4979.37	184.42	175	6180.07	228.89
142	5014.68	185.73	176	6215.38	230.20
143	5050.00	187.04	177	6250.70	231.51
144	5085.31	188.34	178	6286.01	232.82
145	5120.63	189.65	179	6321.33	234.12
146	5155.94	190.96	180	6356.64	235.43
147	5191.26	192.27	181	6391.95	236.74
148	5226.57	193.58	182	6427.27	238.05
149	5261.89	194.88	183	6462.58	239.35
150	5297.20	196.19	184	6497.90	240.66
151	5332.51	197.50	185	6533.21	241.97
152	5367.83	198.81	186	6568.53	243.28
153	5403.14	200.12	187	6603.84	244.59
154	5438.46	201.42	188	6639.16	245.89
155	5473.77	202.73	189	6674.47	247.20
156	5509.09	204.04	190	6709.79	248.51
157	5544.40	205.35	191	6745.10	249.82
158	5579.72	206.66	192	6780.42	251.13
159	5615.03	207.96	193	6815.73	252.43
160	5650.35	209.27	194	6851.05	253.76
161	5685.66	210.58	195	6886.36	255.05
162	5720.98	211.89	196	6921.67	256.36
163	5756.29	213.20	197	6956.99	257.67
164	5791.61	214.50	198	6992.30	257.97

Table 53 *(continued)*

m.³	ft.³	yd.³	m.³	ft.³	yd.³
199	7027.62	260.28	530	18716.77	693.21
200	7062.93	261.59	540	19069.92	706.29
210	7416.51	268.13	550	19423.07	719.37
220	7769.23	287.75	560	19776.21	732.45
230	8122.37	300.83	570	20129.36	745.53
240	8475.52	313.91	580	20482.51	758.61
250	8828.67	326.99	590	20835.65	771.69
260	9181.81	340.07	600	21188.80	784.77
270	9534.96	353.15	610	21541.95	797.85
280	9888.11	366.23	620	21895.09	810.93
290	10241.25	379.31	630	22248.24	824.01
300	10594.40	392.39	640	22601.39	837.09
310	10947.55	405.46	650	22954.53	850.17
320	11300.69	418.54	660	23307.68	863.25
330	11653.84	431.62	670	23660.83	876.33
340	12006.99	444.70	680	24013.97	889.41
350	12360.13	457.78	690	24367.12	902.49
360	12713.28	470.86	700	24720.27	915.57
370	13066.43	483.94	710	25073.41	928.64
380	13419.57	497.02	720	25426.56	941.72
390	13772.72	510.10	730	25779.71	954.80
400	14125.87	523.18	740	26132.85	967.88
410	14479.01	536.26	750	26486.00	980.96
420	14832.16	549.34	760	26839.15	994.04
430	15185.31	562.42	770	27192.29	1007.12
440	15538.45	575.50	780	27545.44	1020.20
450	15891.60	588.58	790	27898.59	1033.28
460	16244.75	601.66	800	28251.73	1046.36
470	16597.89	614.74	810	28604.88	1059.44
480	16951.04	627.82	820	28958.03	1072.52
490	17304.19	640.90	830	29311.17	1085.60
500	17657.33	653.98	840	29664.32	1098.68
510	18010.48	667.05	850	30017.47	1111.76
520	18363.63	680.13	860	30370.61	1124.84

Table 53 *(continued)*

m^3	ft^3	yd^3	m^3	ft^3	yd^3
870	30723.76	1137.92	940	33195.79	1229.47
880	31076.91	1151.00	950	33548.93	1242.55
890	31430.05	1164.08	960	33902.08	1255.63
900	31783.20	1177.16	970	34255.23	1268.71
910	32136.35	1190.23	980	34608.37	1281.79
920	32489.49	1203.31	990	34961.52	1294.87
930	32842.64	1216.39	1000	35314.67	1307.95

TEMPERATURE

Table 54 Degrees Celsius to Degrees Fahrenheit

1 degree celsius = 1.8 degrees fahrenheit

C	F	C	F
0	32.0	32	89.6
1	33.8	33	91.4
2	35.6	34	93.2
3	37.4	35	95.0
4	39.2	36	96.8
5	41.0	37	98.6
6	42.8	38	100.4
7	44.6	39	102.2
8	46.4	40	104.0
9	48.2	41	105.8
10	50.0	42	107.6
11	51.8	43	109.4
12	53.6	44	111.2
13	55.4	45	113.0
14	57.2	46	114.8
15	59.0	47	116.6
16	60.8	48	118.4
17	62.6	49	120.2
18	64.4	50	122.0
19	66.2	51	123.8
20	68.0	52	125.6
21	69.8	53	127.4
22	71.6	54	129.2
23	73.4	55	131.0
24	75.2	56	132.8
25	77.0	57	134.6
26	78.8	58	136.4
27	80.6	59	138.2
28	82.4	60	140.0
29	84.2	61	141.8
30	86.0	62	143.6
31	87.8	63	145.4

Table 54 *(continued)*

C	F	C	F
64	147.2	83	181.4
65	149.0	84	183.2
66	150.8	85	185.0
67	152.6	86	186.8
68	154.4	87	188.6
69	156.2	88	190.4
70	158.0	89	192.2
71	159.8	90	194.0
72	161.6	91	195.8
73	163.4	92	197.6
74	165.2	93	199.4
75	167.0	94	201.2
76	168.8	95	203.0
77	170.6	96	204.8
78	172.4	97	206.6
79	174.2	98	208.4
80	176.0	99	210.2
81	177.8	100	212.0
82	179.6		

	Metric (Degrees Celsius)	Customary (Degrees Fahrenheit)
Water freezing point	0	32
Water boiling point	100	212
Normal body temperature	37	98.6

Part III

APPROXIMATE METRIC CONVERSION FACTORS

Note: Numbers followed by an asterisk (*) are approximate factors. All others are exact.

Length Conversion Tables

Customary to Metric

Unit	Multiply by	To Find
inches	25.4	millimeters
	2.54	centimeters
	.0254	meters
feet	30.48	centimeters
	.3048	meters
	.0003048	kilometers
yard	91.44	centimeters
	.9144	meters
miles	160934.4	centimeters
	1609.344	meters
	1.609344	kilometers
nautical miles	1852.	meters

Metric to Customary

millimeters	.039*	inches
	.003*	feet
centimeters	.394*	inches
	.033*	feet
	.011*	yards
meters	39.370*	inches
	3.281*	feet
	1.094*	yards
	.0006*	miles
	.0005*	nautical miles
kilometers	3280.840	feet
	1093.613	yards
	.621	miles
	.540	nautical miles

Customary to Customary

inches	.083	feet
	.028	yards

Unit	Multiply by	To Find
feet	12	inches
	.333	yards
	.00019	miles
	.00016	nautical miles

Area Conversion Tables

Customary to Metric

Unit	Multiply by	To Find
square inches	645.16	square millimeters
	6.4516	square centimeters
	.00064516	square meters
square feet	929.0304	square centimeters
	0.09290304	square meters
	0.0009*	ares
square yards	8361.2736	square centimeters
	.83612736	square meters
	.00836*	ares
	.00008*	hectares
acres	4046.8564224	square meters
	40.468564224	ares
	.405*	hectares
square miles	2589988.110336	square meters
	25899.881*	ares
	258.999*	hectares
	2.590	square kilometers

Metric to Customary

Unit	Multiply by	To Find
square millimeters	.00155*	square inches
square centimeters	.155*	square inches
	.001*	square feet
	.00012*	square yards

Unit	Multiply by	To Find
square meters	1550.003*	square inches
	10.764*	square feet
	1.196	square yards
	.00025	acres
ares	1076.391*	square feet
	119.599*	square yards
	.0247*	acres
hectares	107639.1*	square feet
	11959.9*	square yards
	2.471	acres
	.00386*	square miles
square kilometers	10763910.*	square feet
	1195990.*	square yards
	247.1054*	acres
	.386*	square miles

Customary to Customary

Unit	Multiply by	To Find
square inches	.007*	square feet
	.00077*	square yards
square feet	144*	square inches
	.111*	square yards
	.000023	acres
square yards	1296	square inches
	9	square feet
	.0002*	acres
	.0000003	square miles
acres	43560	square feet
	4840	square yards
	.0015625	square miles
square miles	27878400	square feet
	3097600	square yards
	640	acres

Weight Conversion Tables

Customary to Metric

Unit	Multiply by	To Find
grains	64.79891	milligrams
	.32399455	carats
	.06479891	grams
	.000065*	kilograms
drams (avoirdupois)	1771.845*	milligrams
	8.860*	carats
	1.772*	grams
	.00177	kilograms
ounces (avoirdupois)	141.747615625	carats
	28.349523125	grams
	.02835*	kilograms
ounces (troy)	155.517384	carats
	31.1034768	grams
	.031*	kilograms
pounds (avoirdupois)	453.59237	grams
	.45359237	kilograms
	.00045	metric tons
pounds (troy)	373.2417216	grams
short tons	907.18474	kilograms
	.90718474	metric tons

Metric to Customary

Unit	Multiply by	To Find
milligrams	.0154*	grains
	.000564*	drams
carats	3.086*	grains
	.113*	drams (avoirdupois)
	.007*	ounces (avoirdupois)
	.0064*	ounces (troy)
grams	15.432*	grains
	.564*	drams (avoirdupois)
	.035*	ounces (avoirdupois)
	.032*	ounces (troy)
	.002*	pounds (avoirdupois)
	.003	pounds (troy)

Unit	Multiply by	To Find
kilograms	15432.36*	grains
	564.3834*	drams (avoirdupois)
	35.274*	ounces (avoirdupois)
	32.151*	ounces (troy)
	2.205*	pounds (avoirdupois)
	2.679*	pounds (troy)
	.001*	short tons
metric tons	2204.623*	pounds (avoirdupois)
	1.102	short tons

Customary to Customary

Unit	Multiply by	To Find
grains	.0366*	drams (avoirdupois)
	.0023*	ounces (avoirdupois)
	.00208*	ounces (troy)
	.00014*	pounds (avoirdupois)
	.00017*	pounds (troy)
drams (avoirdupois)	27.34375	grains
	.0625	ounces (avoirdupois)
	.057	ounces (troy)
	.0039*	pounds (avoirdupois)
	.0047	pounds (troy)
ounces (avoirdupois)	437.5	grains
	16	drams (avoirdupois)
	.911*	ounces (troy)
	.0625	pounds (avoirdupois)
	.076*	pounds (troy)
ounces (troy)	480	grains
	17.554*	drams (avoirdupois)
	1.097*	ounces (avoirdupois)
	.0686*	pounds (avoirdupois)
	.083*	pounds (troy)
pounds (avoirdupois)	7000	grains
	256	drams (avoirdupois)
	16	ounces (avoirdupois)
	14.583*	ounces (troy)

Unit	Multiply by	To Find
	1.215*	pounds (troy)
	.0005	short tons
pounds (troy)	5760.0*	grains
	210.6514*	drams (avoirdupois)
	13.1657*	ounces (avoirdupois)
	12	ounces (troy)
	.823*	pounds (avoirdupois)
short tons	2000	pounds (avoirdupois)

Capacity Conversion Tables – Liquid Measure

Customary to Metric

Unit	Multiply by	To Find
minims	.0616*	milliliters
teaspoons	4.9289*	milliliters
tablespoons	14.787*	milliliters
fluid ounces	29.5735*	milliliters
	.02957*	liters
gills	118.294*	milliliters
	.118*	liters
cups	236.5882365	milliliters
	.02366*	liters
liquid pints	473.176473	milliliters
	.473176473	liters
liquid quarts	946.352946	milliliters
	.946352946	liters
gallons	3785.411784	milliliters
	3.785411784	liters
cubic inches	16.387064	milliliters
	.016387064	liters
	.000016*	cubic meters
cubic feet	28.316846592	liters
	.0284*	cubic meters
cubic yards	764.555*	liters
	.7646*	cubic meters

Metric to Customary

milliliters	16.231*	minims
	.203*	teaspoons
	.0676*	tablespoons
	.0338*	fluid ounces
	.00845*	gills
	.0042*	cups
	.0021*	liquid pints
	.001*	liquid quarts
	.00026*	gallons
	.061*	cubic inches
liters	33.814*	fluid ounces
	6.4535*	gills
	4.227*	cups
	2.113*	liquid pints
	1.057*	liquid quarts
	.0264*	gallons
	61.024*	cubic inches
	.035*	cubic feet
	.0013*	cubic yards
cubic meters	264.172*	gallons
	61023.74	cubic inches
	35.315*	cubic feet
	1.308*	cubic yards

Customary to Customary

minims	.0155	teaspoons
	.004*	tablespoons
	.002	fluid ounces
	.00052*	gills
	.00376	cubic inches
teaspoons	80	minims
	.0333*	tablespoons
	.1667*	fluid ounces

Unit	Multiply by	To Find
	.0417*	gills
	.021*	cups
	.0104*	liquid pints
	.301*	cubic inches
tablespoons	240	minims
	3	teaspoons
	.5	fluid ounces
	.125	gills
	.0625	cups
	.03125	liquid pints
	.90234375	cubic inches
fluid ounces	480	minims
	6	teaspoons
	2	tablespoons
	.25	gills
	.125	cups
	.0625	liquid pints
	.03125	liquid quarts
	.0078125	gallons
	1.8047*	cubic inches
	.001*	cubic feet
gills	1920	minims
	24	teaspoons
	8	tablespoons
	4	fluid ounces
	.5	cups
	.25	liquid pints
	.125	liquid quarts
	.03125	gallons
	7.21875	cubic inches
	.0042*	cubic feet
cups	3840	minims
	48	teaspoons
	16	tablespoons
	4	fluid ounces

Unit	Multiply by	To Find
	2	gills
	.5	liquid pints
	.25	liquid quarts
	.0625	gallons
	14.4375	cubic inches
liquid pints	7680	minims
	16	fluid ounces
	4	gills
	2	cups
	.5	liquid quarts
	.125	gallons
	28.875	cubic inches
	.0167*	cubic feet
liquid quarts	15360	minims
	32	fluid ounces
	8	gills
	4	cups
	2	liquid pints
	.25	gallons
	57.75	cubic inches
	.03342*	cubic feet
gallons	61440	minims
	128	fluid ounces
	32	gills
	16	cups
	8	fluid pints
	4	fluid quarts
	231	cubic inches
	.13368*	cubic feet
cubic inches	265.974*	minims
	3.325*	teaspoons
	1.108*	tablespoons
	.554*	fluid ounces
	.1385*	gills
	.069*	cups

Unit	Multiply by	To Find
	.0346*	liquid pints
	.017*	liquid quarts
	.0043*	gallons
	.0006*	cubic feet
	.00002*	cubic yards
cubic feet	957.5065*	fluid ounces
	239.3766*	gills
	119.6883*	cups
	59.844*	liquid pints
	29.922*	liquid quarts
	7.4805*	gallons
	1728	cubic inches
	.037*	cubic yards
cubic yards	201.974*	gallons
	46656	cubic inches
	27	cubic feet

Capacity Conversion Tables – Dry Measure

Customary to Metric

Unit	Multiply by	To Find
pints	.5506*	liters
	.05506*	dekaliters
quarts	1.1012*	liters
	.1101*	dekaliters
pecks	8.8098*	liters
	.881*	dekaliters
	.0088*	cubic meters
bushels	35.239*	liters
	3.524*	dekaliters
	.035*	cubic meters

Unit	Multiply by	To Find
Metric to Customary		
liters	1.816*	pints
	.908*	quarts
	.1135*	pecks
	.028*	bushels
dekaliters	18.162*	pints
	9.081*	quarts
	1.135*	pecks
	.284*	bushels
	610.237*	cubic inches
	.353*	cubic feet
cubic meters	113.5104*	pecks
	28.376*	bushels
Customary to Customary		
pints	.5	quarts
	.0625	pecks
	.015625	bushels
	33.6003*	cubic inches
	.0194*	cubic feet
quarts	2	pints
	.125	pecks
	.03125	bushels
	67.2006*	cubic inches
	.0389*	cubic feet
pecks	16	pints
	8	quarts
	.25	bushels
	537.605	cubic inches
	.3111*	cubic feet
	.0115*	cubic yards
bushels	64	pints
	32	quarts
	4	pecks

Unit	Multiply by	To Find
	2150.42	cubic inches
	1.244*	cubic feet
	.046*	cubic yards
cubic inches	.02976*	pints
	.01489*	quarts
	.00186*	pecks
	.000465*	bushels
	.00058*	cubic feet
	.00002*	cubic yards
cubic feet	51.428*	pints
	25.714*	quarts
	3.214*	pecks
	.80356*	bushels
	1728	cubic inches
	.037*	cubic yards
cubic yards	86.785*	pecks
	21.696*	bushels
	46656	cubic inches
	27	cubic feet

Volume Conversion Tables

Customary to Metric

Unit	Multiply by	To Find
cubic inches	16.387064	cubic centimeters
	.000016*	cubic meters
	.016387064	liters
cubic feet	28316.8466*	cubic centimeters
	.0283*	cubic meters
	28.316846592	liters
cubic yards	.76455*	cubic meters
	764.554857984	liters

Unit	Multiply by	To Find
Metric to Customary		
cubic centimeters	.061*	cubic inches
	.000035	cubic feet
cubic meters	61023.74	cubic inches
	35.3147*	cubic feet
	1.308*	cubic yards
liters	61.024*	cubic inches
	.035*	cubic feet
	.0013*	cubic yards
Customary to Customary		
cubic inches	.00058*	cubic feet
	.00002*	cubic yards
cubic feet	1728	cubic inches
	.037*	cubic yards
cubic yards	46656	cubic inches
	27	cubic feet

Temperature Conversion Tables

Metric to Customary

Unit	Multiply by	And Add	To Find
Degrees Centigrade	9/5	32	Degrees Fahrenheit

Customary to Metric

Unit	Multiply by	And Subtract	To Find
Degrees Fahrenheit	5/9	32	Degrees Centigrade

Buying Guide

A comparison of Continental and U.S. sizes in clothing

MEN'S CLOTHING

Shirts

United States	14	14½	14½	15	15¼	15¾	16	16½	17	17¼
Continental	35	36	37	38	39	40	41	42	43	44

Shoes

United States	6½	7½	8	9	9	10	10½	11½	12
Continental	39	40	41	42	43	44	45	46	47

Hats

United States	6⅞	7	7⅛	7¼	7⅜	7½	7⅝
Continental	55	56	57	58	59	60	61

Socks

Sizes are the same throughout the world.

General (slacks, suits, etc.)

United States	34	35	36	37	38	39	40	42
Continental	34	36	38	40	42	44	46	48

WOMEN'S CLOTHING

Dresses, skirts, and slacks									
United States	8	10	12	14	16	18			
Continental	38	40	42	44	46	48			
Sweaters and shirts									
United States	10	12	14	16	18	20			
Continental	38	40	42	44	46	48			
Shoes									
United States	4½	5	5½	6	6½	7	7½	8	8½
Continental	35½	36	36½	37	37½	38	38½	39	39½
Stockings									
United States	8½	9	9½	10	10½				
Continental	1	2	3	4	5				

BIBLIOGRAPHY

Kempf, Alfred F., and Thomas J. Richards, *The Metric System Made Simple*; Garden City, NY: Made Simple Books, Doubleday & Company, Inc., 1973.

U. S. Department of Commerce/National Bureau of Standards, *Report to the Congress, A Metric America: A decision whose time has come*; U. S. Government Printing Office, 1971.

U. S. Department of Commerce/National Bureau of Standards, *Some References on Metric Information*, Special Publication 389; U. S. Government Printing Office, 1974.

U. S. Department of Commerce/National Bureau of Standards, *U. S. Metric Study Interim Report, Commercial Weights and Measures*, Special Publication 345-3; U. S. Government Printing Office, 1971.

U. S. Department of Commerce/National Bureau of Standards, *Units of Weight and Measure, International (Metric) & U. S. Customary*, Miscellaneous Publication 286; U. S. Government Printing Office, 1967.

U. S. Department of Commerce/National Bureau of Standards, *What About Metric?*; U. S. Government Printing Office, 1974.

United States Metric Board, *U. S. Metric Board Information Kit*; 1815 North Lynn Street, Arlington, VA 22209, 1975.

24B